FALLING FOR IT:
THE APPRENTICESHIP

FALLING FOR IT:
THE APPRENTICESHIP

Geoff Surtees

FARMING PRESS

First published 1995

ISBN 0 85236 299 4

A catalogue record for this book is available
from the British Library

Published by Farming Press Books
Wharfedale Road, Ipswich IP1 4LG, United Kingdom

Distributed in North America
by Diamond Farm Enterprises,
Box 537, Alexandria Bay, NY 13607, USA

Cover design by Mark Beesley
Cover photograph by kind permission of the Forestry Commission, Edinburgh
Text illustrations by Pete Wilford
Typeset by Galleon Typesetting, Ipswich
Printed and bound in Great Britain by Biddles Ltd, Guildford and King's Lynn

Introduction

Born sucking a piece of grass, I was programmed for a rustic role. Forestry chose me when I left school and, after a compatible and lasting relationship, I have been left a legacy. This "inheritance", I hasten to add, does not show on bank statements; it comes in other forms. But, regrets I do not harbour; I am a wealthy man.

With an enquiring mind, an alert eye and a pricked ear, I have worked through the seasons, harvesting the rewards of my vocation; and after more than three decades in the woods – I am rich with tales to tell!

Particularly vivid in my mind are the first three years of my working life, when I was introduced to forestry and the rural ways on an estate in Northumberland.

I remember the work, its interest and challenge; the estate, with its lingering traces of feudalism, and the wildlife with which we co-existed – but mostly, I recall the characters I was privileged to meet – all of them, true countrymen.

Often, when recounting incidents of interest, I have been asked to explain exactly what I mean by "true countrymen". This seemingly reasonable question has repeatedly left me sinking in a peat bog of my own making – alas, there is no simple definition. The complex diversity of this sadly dwindling breed is so great that qualification is virtually impossible. At least it was until recently.

A countrywoman invited a city-dwelling male friend to spend Christmas with her in the tranquillity that only Northumberland can offer. It would be peaceful (she knew the jet fighters were grounded to have their silencers polished), and a complete change for her guest. The preparations were almost over − but the final touch was missing.

"I'm just going out to get some berried holly", said the hostess to her guest as she donned her coat. This apparently simple announcement drew her friend's eyes away from the television, and with a thoughtful look he asked the question that, finally, has given me the elusive answer I sought: A "true countryman" knows where to dig for berried holly.

It is guilt at the indecent hoard of rustic experiences I have amassed in the course of my life that has shamed me into sharing them. I can no longer sleep with the burden!

This is the story of my first three years at work; and, if you enjoy reading it one-tenth as much as I enjoyed living it, the effort in writing it will have been worthwhile. Happy reading.

This book is dedicated to a wonderful cast of country-folk,
with a special thought for Don

Chapter One

Early Days

I was a chronic countryman by the time I left school, and a disappointment to my mother. Mother wanted me to work in a bank in the hope that someday I would become a bank manager. Why she thought I had the qualities to rise to that elevated position I will never understand; I did not pull wings off flies! Fortunately Northumberland had other plans.

It was 1947, and I was four, when my mother and I moved into a large and impressive, red-brick bungalow, on the outskirts of a small village in north-west Durham. Surrounded by fields the dwelling proudly stood atop a steep hill, overlooking Northumberland's rural splendour across the Derwent valley. The bungalow may have been home, but the Northumbrian countryside was my playground, and it was to have a profound and lasting influence.

I spent my youth detesting school, and devoted every hour free of it to the woods and fields that lured me. I walked miles learning to read the signs of the countryside. I caught rabbits, climbed trees and tickled trout. I knew every fox earth, badger sett and rabbit hole. I could stalk the shy roe deer, locate the precocious red squirrels and elude gamekeepers with ease. I was as wild as the countryside – and hardly cut out for a career in banking!

Long before I had a chance to think seriously of what I would do for a living, necessity had introduced me to the practice of earning money. Short of pocket-money, I had to supplement my allowance, and I was surrounded by a solution: rabbits abounded, and catching them provided a steady income in my early years, until disaster struck in 1955 – myxomatosis!

Rabbits were a pest, and I had killed them, but I was saddened by their plight. It was sickening to watch the once irrepressible creatures suffer a mass, lingering death. The dying ate among the dead as the barbaric myxomatosis waited a persistent ten days to claim its victim. Blind and deaf, with grotesquely swollen, puss-oozing heads, the rabbits ate until the end. In the early stages the stricken, deprived of sight and sound, were still alert to the heavy footsteps of a careless approach; and never will I erase the memory of a startled creature pathetically fleeing, to die in a thud of collision with the bottom of a dry stone wall.

It took just a few disgusting weeks to wipe the rabbit out. A weird stillness, alien to the countryside, settled through the stench of rotting flesh. The fields and woods that once rippled with life were as dead as their former animators.

I respected the humble rabbit: a true natural survivor that, despite the interest of almost every predator, not only stood its ground, but increased to plague proportions. This, achieved by a species with a single line of defence: mastery of the art of procreation. They did not deserve their fate.

After the decimation of the rabbit I turned to local farmers for a source of money, and could usually find work. Hay and harvest time were most enjoyable, probably be-cause of the better weather and the fact that it was more of an occasion than work. Potato picking, and singling (thinning) turnips, never seemed to pay in proportion to effort and discomfort, but worst by far was stone picking, a job fit only for the corrective treatment of offenders.

I never had the usual boyhood desires to be a bus or engine driver; only two possibilities had occurred. Surprisingly, one was farming and the other gamekeeping. The former held many attractions, especially when I recalled the more enjoyable times, but some of the agricultural tasks were nothing more than a boring, unpleasant grind – or worse in the case of stone picking.

Gamekeeping had slightly more than equal appeal. It involved wildlife and woodland; it offered a greater challenge and freedom, but there were doubts. Gamekeepers in those days did too much forelock tugging, and wasted far too much time in futile pursuit of suspected young poachers for my liking.

Happy about neither option, I pondered long, before my vocation was discovered when talking to an older friend.

"Have you thought about forestry? Ivan Hunter's been working in the woods for over a year, and he likes it. Have a word with him."

Thought about it? No, I had not. It had never even entered my head . . . Yet why not? It offered everything. It fitted – perfectly.

The choice was made – forestry. Later discussions with Ivan merely reinforced my enthusiasm, and finally confirmed the decision. All I had to do was wait until my day of freedom from school. It came when I was 16, in 1959.

Enquiries into the chosen career revealed two likelihoods. Firstly, personal transport would be needed, as "frequent" is hardly the word to describe the movements of the rural bus. And, secondly, I would almost certainly have to travel into Northumberland where acres of timber stood.

Acquired in a straight swap for four five-pound notes, a Bantam motorcycle solved the mobility problem. I soon learned to ride it, and discovered its hidden potential. Given ideal conditions, like over the edge of a precipice with a strong following wind, this machine would reach an air-shattering 55 mph – and maintain it for almost one minute before threatening to seize up! Nevertheless, with sympathetic understanding, it would serve its purpose.

Northumberland was the county of forestry; it was there that it happened on a grand scale, and the obvious destination of my first letter of application was the Forestry Commission.

Northumberland fits conveniently into the top right-hand corner of the map of England. A county of sparsely populated rural expanse, it guards volumes of history and secrets. Northumberland has more land under afforestation than any other English county, and that is because of the Forestry Commission. Formed in 1919, the Commission had one specific objective: to promote forestry and expand timber production – a task that is simple in theory, but formidable in practice. Devastated by the demands of the war, the nation's timber resources had to be replaced, and increased to provide for future needs.

The Forestry Commission's Kielder Forest is the largest man-made spruce forest in Europe; omit the word "spruce" and it is open to argument. Today the forest spreads into Cumbria, but the first planting took place in 1926–27 on Northumbrian soil. Work continued and by 1962 some 32,000 hectares, or 125 square miles, of Northumberland had been afforested in this one area. The last forest census was taken in 1982 when the trees of Kielder occupied 38,000 hectares – yet this represents only half of the total

76,000 hectares, almost 300 square miles, of England's finest county, carpeted by trees.

At last it had arrived: a brown envelope marked Forestry Commission, and addressed to Mr G. Surtees lay face up on the doormat. It was the reply to my initial enquiry to head office requesting details of a career with the Commission. The letter thanked me for my interest, congratulated me on my choice of occupation, and suggested I write to the Commission Districts; an enclosed booklet answered every conceivable question on forestry as a career. The plan unfolded in detail: two years practical with the Commission, a similar period attending a forester training school, and I would emerge a qualified forester – straightforward.

I lost no time in writing to the nearest Commission District, and slipping the envelope into the post box, I wondered where and when my interview would be.

But it did not happen like that – disappointment came in reply. The letter explained, apologetically, that the present policy was not to take on any more employees, especially the inexperienced, for the foreseeable future. Other Districts were similarly affected, and it was suggested I contact the private estates on the enclosed list. The letter closed wishing me success.

The plan had not anticipated a negative answer. The Commission with all those acres under trees was bound to be short-handed . . . How wrong I had been! My first two years with the Commission were not to be.

Born an optimist, and undiscouraged, I scanned the list of private estates and decided on a strategy. Beginning with the nearest to home, I would visit each estate and present myself directly to the forester. In theory this short-circuited the letter of application and an interview was guaranteed; and, assuming it would be more difficult for a man to say no to an eager youth who had taken the trouble to call in, it seemed good tactics.

First on the list was the local estate.

"Good morning, Colonel Masson. I'm Geoff Surtees," I said, fighting my nervousness. "I'm sorry to trouble you, but I was wondering if you have any vacancies in the woods?"

"Yes, as a matter of fact I do," came the swift reply that immediately boosted my hopes. "Unfortunately, we need a skilled man, and you're straight from school. Have you anywhere else to try?"

"Yes, three," I replied, saturated with disappointment.

"Well, try the others, but if you don't find what you're looking for, call back."

The parting words offered hope I could hold in reserve, but as the bike took me home something began to tell me that all might not be as simple as I had first thought.

My next visit produced the same result: again a man was required, but he had to have experience. I now accepted I had a problem, but press on regardless I would. The third call, I decided, would be lucky, and morale built up by positive thinking on the journey was high as I posed the question.

"Aye, lad, we do need a man, but you aren't experienced," was the firm reply. No! Not again, I thought, as I crashed back to earth. Strange, everywhere I went men were wanted, but they had to have experience; nobody was prepared to start a trainee – strange indeed.

Blessed with rose-tinted lenses, my eyes pictured the brighter view, but the gift was perhaps overgenerous, and at the expense of another quality which I lacked – patience – I had a very short temper. I knew it, and tried to control it, but nothing kindled my fire more quickly than a lack of common sense. The illogic of the forestry employment policy was beginning to annoy, but I assured myself that the fourth, and last, chance would be different and rewarding. The forester was Mr J. Shaw.

"Good afternoon. Mr Shaw?"

"Yes, lad. What can I do for you?"

"The name's Geoff Surtees. I just happened to be passing, and I thought I would call in to see if you have a vacancy in the woods."

"Indeed we have. Trouble is, we're looking for . . ."

I did not hear the end of the sentence; I did not need to as I had heard it all before, and could not believe that I was on the receiving end of those same words yet again. The fuse had burnt; the temper flared; there was no job here – I had lost; lost all but my bottled frustration.

"Hell! Not again. If I've heard that once I've heard it a dozen times," I exploded, with a pinch of exaggeration. "Everywhere I've been, men are wanted – provided they have experience. But nobody will take on a trainee. If skilled men are needed everywhere, then it's obvious there's a shortage, and the only solution is to train people. What a stupid situation!" I protested – loudly.

My once-prospective employer may have been taken aback by my outburst. He just stood there, his face devoid of expression, making no attempt to speak. I was in full cry and continued without interruption.

"And another thing: if nobody is prepared to train people, what happens when the present men retire? I'll tell you. There won't be anyone at all working in the woods. What a shortsighted attitude. If that's how foresters think, I don't want to work in the woods!"

I had voiced my thoughts, and vented my wrath; now I faced the desolation of the journey home – and a grovelling return to the local estate.

"Thanks anyway," I growled, for some peculiar reason

remembering my manners as I turned towards the bike. I grabbed the handlebars of the Bantam, and was about to throw my leg over the seat when a voice from behind ordered, "Come back here a minute, son!"

I turned round to see Mr Shaw standing where I had left him. A man in his thirties, tall and of medium build, with a handsome rubicund face topped with greying hair – he beckoned.

Head bowed in shame I trudged towards him, certain to receive at least a lecture to the effect that being cheeky was not the best way to get a job. I deserved a reprimand, and stood before him prepared. Pale blue eyes fixed on mine from a still-expressionless face. The gaze was cool and lasting; the lull before the storm. I tensed as he was about to speak . . .

"What you've just said, son . . . is perfectly true. Can you start on Tuesday?"

First Day

Quarter past seven on the morning of 17 November 1959 found me astride the Bantam, kicking life into the two-stroke engine. Bait bag on back, and with both excitement and apprehension riding pillion, I opened the throttle and revved away into the chill of the morning air.

I arrived ten minutes early, pulling up in the estate yard beside a smiling Mr Shaw.

"Morning, son, put your bike in the stable," he beamed, indicating a building with an outstretched arm. I rode the Bantam through the door and parked it beside a large rotovator. Looking round inside the stone walls that had once housed horses, I was amazed by the number and variety of the present contents – tools. Spades, rakes, axes, sickles, slash knives and hoes, to name but a few, hung from nails along the walls, their steel polished by regular use. Wheelbarrows, fencing wire and strange wooden objects filled up the floor space.

By the time I walked out of the door towards my new boss the old hands were coming into the yard on various modes of transport.

"Guid morrrnin' laddie," I was greeted in a broad Scots accent by a pipe-smoking, contented old chap as he passed on foot.

"Good morning," I nodded in reply, as another older man dismounted an outsize lady's bicycle.

Yet another arrival came on a moped pursued by a trail of thick blue smoke. Fascinating was the figure the moped brought in. Above a pair of shining black boots, the trouser bottoms of "bib and brace" overalls were tied with baler twine round a thin pair of ankles. Generously cut, the overalls fitted only where they managed to touch the short, slight frame of their occupier. A painfully tight jacket, grey and pin-striped, corsetted his upper torso, and the outfit was completed by a cap that could easily have doubled as a dustbin lid! I was not to know that I would never see that cap removed, and to this day I do not know whether the ears to the sides of his thin-featured face protruded in support of the cap or protruded because of it!

Of course I knew nobody who worked there, but had hoped to find someone nearer to my own age, so I was pleased to see a much younger man ride in on a Norton-powered combination. In place of the sidecar was a large wooden box.

Mr Shaw was standing beside the rear wheel of a Fordson Major tractor, garaged in an old cart shed. As I approached the open doorway I could see a man swinging the starting handle. The engine spluttered hesitantly before bursting into healthy song.

"It is Geoff, isn't it?" queried Mr Shaw.

"Yes, it is," I nodded.

"Good. Only the best people are given that name son," replied my boss, telling me what his "J" stood for.

Turning to the man who had started the tractor, Mr Shaw introduced me. "Geoff, this is Will; he's the foreman and tractorman, and today you can help him – he's going peeling."

Will and I exchanged pleasantries as the engine warmed up sufficiently to switch fuel taps from petrol to paraffin. Ready to go, Will climbed onto the tractor seat, indicating that I take position on the heavy winch, firmly fixed by huge bolts to the rear of the tractor. A thick, folded hessian sack cushioned my rear as we set off into my first day of work – peeling, whatever that was.

Pulling out of the yard, Will flung out his left hand, pointing to a large enclosure surrounded by a high wire-netting fence. "That's the nursery; we'll all be in there soon," he shouted above the noise of the engine.

The nursery looked like an enormous allotment, with countless trees, row after row, instead of vegetables. Not all estates had nurseries, and I was pleased to think that I would gain the experience of growing trees from seed.

Turning right onto the North Drive we followed the gentle fall of the road as it carefully tunnelled through an avenue of stately trees. The fresh smells of the early hours mixed with the occasional pleasant whiff of exhausted paraffin as we sedately travelled over a cattle-grid, and out into open parkland.

Mature limes and oaks dotted the pasture, proudly displaying spreading limbs and branches bereft of leaves in preparation for winter's rest. The stillness of the morning was broken as the characteristic white heads of Hereford cattle loomed from the background, bouncing across the park on stampeding bodies as we were charged in mistake for a delivery of hay. The "Big House", in all its grandeur, stood aloft and aloof on our right, overseeing the parklands and beyond from a position of vantage.

Another cattle-grid marked the end of the park, and once again we entered a tree- and shrub-lined drive. Rhododendrons flourished, guiding us towards an old well that sadly lay in dereliction. Ornate stonework, disappearing beneath enveloping moss, hinted at its previous importance. A left-handed bend took us onto a straight that gradually descended through unbroken mixed woodland to the bottom gateway that was only just discernible in the distance. A pair of tiny cottages, one on either side of the drive, and a house further on took me by surprise as they leapt from the cover of the trees that had hidden them until we were opposite.

Through the gate by the North Lodge we crossed the main road, and negotiated a short forest track, known as a ride, before bearing right along a byroad. On our left an evergreen plantation showed signs of recent activity: freshly cut branches littered the forest floor, and heaps of newly cut timber waited ahead. Will eased the throttle as we neared the stacked wood, pulled in and wrenched on the hand-brake.

"This is it, lad. Not a bad job, peelin'!" Will happily stated as he dismounted and, pointing to the neat stacks of timber, explained, "These are all pit props, Geoff, and they range from two feet in length in this stack, to six feet with a five inch top diameter over there. We have to peel, or debark, them all, and this is the peeler."

I had seen a machine similar to the one in front of me on some of my younger excursions, and I knew what it did, but did not know it was termed peeling.

The peeler

Will manoeuvred the tractor into position and handed me a belt which I placed over the lined-up pulleys of both machines. Reverse gear tightened the belt, and the hand-brake was applied. After we had a brief break for a cup of tea and a sandwich, the pulley drive was engaged.

The peeler was nothing more than a heavy steel saucer-shaped flywheel, mounted vertically to the side of a small steel table. Blades like those of wood planes presented their sharp edges through the bottom of the "saucer" which faced the table. The operation was simple in principle. Held at one end, a prop was rested on the peeler's table and turned against the rotating blades until the bark along half its length had been removed in shavings. The prop was turned round and the process repeated, giving a finished article that was both easy to handle and quick to dry out.

My first job was the more menial one of handing Will the unpeeled props and stacking the peeled ones. The two foot props provided the only "excitement". Unlike the longer ones they had to be sorted according to top diameter: two to two-and-a-half inch top onto one stack, and two-and-a-half to three inch on another.

Soon I was hoping that Will would tire, and allow me to try my hand. Peeling seemed easy enough, and it was satisfying to see a prop transformed from a rough dark object to a smooth white one amid a shower of shavings. My wish had not been realised by the time Will turned off the tractor fuel tap and, soon starved of paraffin, the engine fell silent. It was twelve o'clock: bait time.

Sitting down on a sack with my back to a tree, I poured out a flask-cap of tea, and set about the contents of my bait box, congratulating myself on my choice of occupation; everything was right and as it should be. The midday autumn sun fired shafts of warming light through chinks in the canopy, banishing the morning's chill. The air was still, laden with the fresh clean smell of rosin. Nothing moved; the peace was perfect; we could have been the only two people on earth, until the tranquillity was shattered – and my mental excursion brought to an abrupt end – by the clanking churns of a milk wagon as it passed on the nearby road.

Will had finished eating and was engaged in the more serious business of refuelling his pipe with rubbed-out chunks of Condor Bar. In every respect Will was average: his height, his build, his looks and his face all defied description because of their normality.

Joining him in his quest for nicotine I lit a Woodbine, inhaled deeply and asked, "How many men work in the woods, Will?"

"Er . . . eight, includin' yourself," came the reply, after thought.

"A Scotsman spoke to me in the yard and two other men arrived, one on a push-bike and the other on a moped – are they woodmen?"

13

"Yea, they are. The Scot is old Don; he's the sawyer. The one on the bike is Frank, and the moped belongs to Johnny, or 'The Professor' as we call him. Knows about things scientific, does Johnny," said Will, his face breaking into a broad grin, allowing smoke to curl from the corners of his mouth.

"The younger chap on the Norton?"

"Oh yes, that's Tom; he does the fencin'. And there's Ron and Jake as well. Int'restin' feller is Jake. He used to be a horseman, and has an uncanny knack with animals. He's due for retirement soon. But don't worry lad, you'll meet everybody in a couple of days. We'll all be in the nursery on Thursday."

Everyone, at some stage in their life, makes a statement that they live to regret, and I was no exception: "Oh good," I said innocently. "That should be interesting."

My comment was greeted by an immediate spluttering cough, which expelled Will's mouthful of smoke in a series of uncontrolled bursts. He rose slowly, shaking his head and fixed his eyes on mine with a look which I shall never forget. There was an uneasy combination of sympathy and dismay about his countenance; an expression of understanding, yet disbelief. The look spoke louder than words: Will knew something that I did not.

The tractor was restarted, and work resumed. The afternoon unfolded as a carbon copy of the morning. My hidden desires were ignored: Will showed no sign of offering me a change of roles, and I decided to content myself with my subservient lot.

The stacks of unpeeled props decreased while the heaps of their clean white counterparts minus bark grew in proportion. The afternoon flew, and in fading light we retraced our wheeltracks back to the yard, leaving behind only enough peeling for one day.

"Night, Will," I said as I mounted the bike.

"Yea, night, Geoff. Go straight to the wood in the mornin' and we'll finish the peelin'. Then it's the nursery – you'll find that int'restin'!"

I detected a hint of undisguised irony in his last few words, and the journey home gave me time to think, and suspect, that all was not well in the nursery.

Day Two

Back at the peeler the following morning, I took advantage of the time I had before Will arrived to have a look around the wood. In the past I had spent countless hours exploring evergreen plantations, but it was the wildlife that drew me then, not the trees. Now I had a different focus of attention.

As far as I knew there were two types of tree: the deciduous and the evergreen. I could name the common deciduous varieties, but evergreens posed a problem with a difference – the difference being the lack of it – they all looked the same to me. The conical shape, the straight, single trunk and branches bearing needles that were green all year. However, I was aware of one so-called evergreen that for some strange reason renounced its title each autumn by turning yellow and shedding its needles, replacing them in spring with delicate, pale-green growth.

The sound of the tractor engine carried clearly through the freshness. By the time Will had parked the tractor by the peeler I had decided that there were two types of tree in the wood: one a true evergreen and the other a false one that lost its needles.

"Mornin', Geoff. Chilly a bit," Will observed, handing me the belt and holding his hands over the exhaust.

"Morning, Will, certainly is," I agreed, placing the belt over the pulleys and hoping that today it would be my turn to peel. Alas, it did not happen, and I carried on in my labouring capacity.

Will and I were having our morning bait when the puzzle of the deciduous "evergreen" sprang to mind. "Will," I asked, "why have some of the trees in here shed their needles? I thought they were supposed to be evergreen."

" 'Evergreen' is a bad word, Geoff," Will replied. "In the woods we don't use it. Instead of 'deciduous' and 'evergreen', we commonly say 'hardwood' and 'softwood'. But the terms 'broadleaves' and 'conifers' are better still. In this wood, apart from the odd birch that shouldn't be there, all the trees are conifers, and the ones that have lost their needles are larches, the only deciduous conifer."

Useful information, I thought; the larches were quite distinctive with their lack of foliage, and I was wondering what the other type of tree was when Will continued.

"Don't ask me why, it doesn't often happen, but in here they've planted six different trees. Know any of them?"

I had been told to expect a few tricks in the first week or so, and this had to be one, but despite the forewarning I was shaken. I stopped drinking and frantically scanned the trees for some indication of difference . . . There was none, there was only the larch and the green one — and I was being asked if I knew any of the six!

"N-No," I spluttered, "I only know some broadwoods — er — I mean hardlea . . . No damn it! Oaks, beech, sycamore and trees like that," I answered, making a complete mess of my newly learned terminology.

Will sensed my plight and came to the rescue with words of comfort. "There are only three species in here, Geoff. And don't worry, you'll soon get to know them."

"Only three?" I mumbled, taking small consolation from the reduction.

"Yes, three. There's the larch that you know, and that tree you have your back against is a spruce. Look at the bark. It's smooth compared with the larch, and the needles are single and evenly spaced along the branches. This tree behind me is a pine, and you'll notice it has rough bark at the butt, and the needles grow in tufts."

"I see," I said unconvincingly, as we got up to resume work.

For the rest of the morning I completely forgot my wish to take a turn on the peeler. Carrying and stacking props did not engage the mind, and I was free to concentrate on the trees, searching for the elusive differences Will had pointed out. Gradually, and mercifully, the mystery unfolded. I began to detect dissimilarities. I carefully examined the three trees, comparing the bark and needles. I looked closely at the props I was handling, and watched the peeler reveal the different colours of the inner bark. By lunchtime I had mastered the subject.

Bait box empty, I was bursting to demonstrate my new-found ability, and, pointing to three trees in turn, announced, "That is a spruce. That is a pine. And that one is a larch!"

"Good lad!" Will congratulated. "Told you it wasn't hard."

Smug with satisfaction, I awarded myself a Woodbine, and sat back against my spruce to enjoy a well-deserved smoke. "You had me worried when you joked about six different trees being in here, Will. Apart from the larch, the others looked the same, but now I know all three."

"There *are* six, Geoff," came the stunning reply that brought me bolt upright in disbelief.

"Six? You showed me three!"

17

"Yea — three species, but there's two varieties of each in this wood."

My sense of achievement was shattered. I had spent all morning valiantly wrestling with the problem of discerning the species, only to find I had accomplished a mere half of the task — and the easy half at that. The species were reasonably distinct now, but the thought that there were two types of larch, spruce and pine was frightening.

"The spruce you pointed out, Geoff; look at it. A Norway spruce. Smooth, dark-red bark, and dark-green needles. Now look at the one you're sitting beside, a Sitka spruce. Greener bark, and the needles are paler and have a distinct bluish tint, especially on the underside."

Will paused, allowing time for the words to sink in.

"That pine nearest the peeler is a Corsican, and the one behind me is a Scots. The Corsican has longer and greener needles than the Scots, but the most obvious difference is in the bark. Look at the bark higher up: the Scots has unmistakable pinkish bark."

What Will said was true. I could see it, and began to wonder why the obvious had remained hidden until pointed out. The bewilderment receded, and I gained in confidence as Will continued.

"The larches are the hardest to tell apart, Geoff. The one you named is a European, and the one behind the tractor is a Japanese. The easiest way with larches is to look at the newest growth at the ends of the branches. Look at the European and you'll see it's a pale straw colour, but the Jap is a definite red."

Yes; indeed they were. Will had packed a lot of information into a short lesson, and the brain struggled to take it in.

"You can peel this afternoon," Will offered, stopping my thoughts with his words.

18

I did not wait to be told a second time. Prop in hand I stood waiting at the peeler before Will had turned the petrol tap on. At last my turn had come, and I intended to make full use of it. Peeling looked straightforward, and I was determined to do a good job. As Will peeled the props clean of bark, I had noticed that he took care to avoid taking too much off the thin end. As far as I could see the only other thing that mattered was that the fast-moving blades were kept free of knuckles and blood. The prospect of failing and being relieved of my promotion seemed unlikely, but I admit to apprehension as the blades began to turn. I need not have worried: the job was easy. Soon I began thoroughly to enjoy the work as I shaved the props of their bark.

"Three o'clock – pipe time!" Will shouted, as he disengaged the pulley.

As we took our seats it occurred to me I had not thought of the trees while I had been peeling – panic. Could I remember their names?

With a sense of frantic urgency I cast my eyes from tree to tree, wracking the brain to recall the vital information, the differences – what were they? The tension relaxed in stages as one by one the details were rediscovered. However . . .

"Err, Will."

"Yea?"

"In this wood there are European and Japanese larch, Scots and Corsican pine and Sitka and Norway spruce."

"That's exactly right, lad," Will assured with a smile.

"And the differences. The larches: pale shoots against red. The pines: longer, greener needles against reddish bark higher up the tree. The spruce: red bark and dark needles against needles blue on the underside."

"Well done, lad. You've remembered it all; not bad for

one day. You'll have no problems now."

"Afraid I have, Will."

"What's that?"

"I've forgotten which is which!"

Will soon refreshed my memory, and I rode home contented. Not only had I experienced peeling, I had learned to identify six conifers – a lesson I would never forget!

Chapter Two

The Nursery

Its throttle wide open, I pleaded with the Bantam to deliver all the power its 125 cc could muster. I was late. The road climbed hills on the way to work, which did not help to make up the time I had lost earlier, coaxing a reluctant engine to start.

Blasting down the South Drive and into the yard I shot through the open stable door, braked hard and slid to a halt beside the rotovator. One minute to eight – I had made it! I could see Will, and parts of other bodies, standing in the harness room that served as a rendezvous point and bait room when working in the nursery.

"Morning!" I hailed, as I walked through the open door. "Not bad for November."

A cacophony of grunts, groans and coughs was the response of characters, diverse, as they stood with their backs to the wall. All wore caps and pained expressions, their eyes staring vacantly ahead. Teeth clenched pipes, mouths puffed smoke, but not a word passed the lips. An atmosphere of gloom hung as heavily as the dense blue cloud from the Condor Bar that most seemed to smoke.

"This is Geoff, our new starter," Will announced, placing

his hand on my back. "He thinks the nursery will be int'restin'." The final word "interesting" was accentuated by a hefty slap between the shoulder blades that rocked me onto my toes.

I could feel the six pairs of eyes as they turned and fixed me with glares of disgust, pity, horror and amazement. The silence hurt.

Fortunately, I was saved from my embarrassment by Mr Shaw, who swung open the door and strode in. "Right men, let's be having you! We have 25 acres to plant Next Door, so we'll need 50,000 trees lifted; let's get on with it."

The words were laced with an enthusiasm that should have inspired a dead cat to purr. Alas, they fell on rocky ground, and were met with an equally stony silence. We filed out of the harness room, selecting spades from their nails on the stable wall on our way through. Young though I was, and full of the associated exuberance, I was not immune to the prevailing air of despondency as I brought up the rear of a procession of men trudging wearily toward their doom. The message was becoming clear: nobody liked the nursery!

The nursery occupied some three acres – an awe-inspiring expanse when I looked at the spade in my hand. It was divided into three plots or breaks of equal width; breaks one and two were separated by a wide grass road, breaks two and three by a narrow grass path. There were trees everywhere, those in the seedbeds standing only a couple of inches high, those in rows, up to fifteen inches, and growing at two to three inch spacings in rows eleven inches apart. It was horribly obvious that the lifting of 50,000 trees was not going to make much of an impression.

We split into two gangs. Will, Don, Ron and I left the others and went further up the break. The trees had to be sorted and tied into bundles of 50 after lifting, and it was decided that Don and I would do the spadework. Grasping his spade in a pair of horny hands old Don began to show me how to ease the trees from the ground without damaging the roots. The job only required a little care and common sense; nevertheless Don appeared to derive great satisfaction from sharing his knowledge with me. I was not to know that this was the beginning of a working relationship for which I am indebted to this day.

In his prime Don must have been physically impressive and exceptionally strong. Even in his sixties there was only the slightest wilting of his large, muscular frame. Smiling blue eyes beamed from a cheerful, angular face, which was unusually pale for someone who had spent a lifetime working out of doors.

Don and I started at opposite ends of a row. Working toward each other, we lifted and laid the trees behind for Will and Ron to bundle. Except for a brief break at nine o'clock for bait we worked without pause and with little conversation throughout the morning.

Twelve o'clock arrived like a fresh, warm breeze. The reluctant plod that had taken the men into the nursery was exchanged for a sprightly pace as a happy band retraced their steps through the gate. A further rejuvenation of their souls was apparent as they fled in the direction of their homes, or the transport that would take them there. Most of the men lived within a convenient distance of the nursery and took the opportunity to go home for their lunch. In the harness room I had the company only of Ron and Jake.

As we sat eating our sandwiches, I noticed we were alike in one respect: none of us wore the bib and brace overalls accepted as standard by the others. There the similarity ended. Ron, I guessed, was nearing 50, and was nondescript for the same reasons as Will, although I suspected that little hair grew beneath his cap.

Conversation was light, and from the softly spoken one or two word contributions made by Jake, I detected an accent that could only have been acquired in the more northerly expanses of Northumberland. Jake was different, and special. He was a tall man of wiry build, and his head and shoulders bowed their acknowledgment of his 64 years. He wore shepherds' boots with their distinctive turned-up toes. Faded brown corduroy trousers hung from braces, and were tied below the knee with baler twine. Under a well-worn donkey jacket an unbuttoned collarless shirt revealed a mass of hair on his chest. Gnarled hands and his craggy countenance, etched by the weather, bore testimony to a lifetime on the land. A strong sandy moustache sprouted from below his long sharp nose, hiding his mouth as it drooped astride the blunt point of a determined chin. High cheekbones sported tufts of whiskers spared by the razor. A pair of dense, vigorous eyebrows supported the peak of a tattered cord cap. They bristled intimidatingly, yet sheltered a pair of deep-sunken, pale blue eyes, that twinkled peacefully in contrast below.

Jake did not end there; he had more than the eye could see, and I had not been in his company long before I felt it. There was an air about his person. He radiated a warmth, a strange, almost hypnotic influence that soothed and made me at ease in his presence. It was odd, bordering on eerie, but somehow I would have found it impossible to argue with Jake, this uncomplicated man of peace, calm and gentility. He could not harm a wood ant. I found it comforting to know that I was not alone in noticing Jake's vibes; everyone else did, and it was suspected by some that this explained his unique rapport with animals. Apparently animals would do anything for Jake.

Gentle Jake, as I decided to call him, had at his side what could only have been a very faithful companion. It was an old tin flask, and judging by its appearance it must have shared his every day of working life. Everyone else had long since converted to the vacuum flask with all its inherent advantages, but not Jake. Jake was, for some obscure reason, content to partake of the lukewarm liquid held by his primitive friend. Like its owner, the flask had character. A mosaic of indentations decorated the whole of its surface area, the bruises of its years in service, and the original handle had been superseded by a piece of fencing wire. However, today there was a problem. Jake had noticed an arc of moisture where his flask had stood, and with the vessel held high he examined the base from where the contents had started to weep.

*   *   *

At the end of the day, cleaning our spades, Will gave me a long hard look, and asked, "Did you find that int'restin' then, Geoff?"

"No, not interesting, but it doesn't seem all that bad," I answered truthfully. "Why does everyone hate it so much?"

Resting his hand on my shoulder Will turned away and shook his head patiently before looking me straight in the eye. "Imagine a hot, but wet, summer. The ground has been regularly soaked keepin' us out of the nursery for weeks. Then, one day after a fine spell, Jeff comes round with a huge grin on his face and says, 'Right, men, nursery tomorrow.'. . . Now, try to imagine the weeds. They're standin' lush and knee-high – waitin' to meet you at the bloody gate. You fight your way through to find you can't see a single tree for the bloody rotten things. When it gets like that, son, you can spend damn weeks crawlin' about on your bloody knees pullin' the buggers out by hand – out of all three acres. You'll find that really bloody int'restin', son!"

Will's excitement had grown with every word, and his face was quite flushed as he put the finishing touches to his graphic illustration. With the picture clearly painted in my mind, I felt sick at my youthful naivety. Those words I would live to regret – time and time again!

Planting

I had no illusions about the nursery by the time the lifting was finished. Although brief, my education there had been sufficient, and I was thankful to return to the freedom of the woods.

Next Door, I had learned, was a neighbouring estate acquired by our landowner just two years earlier. The previous owner had shown little interest in forestry, so most of the woods were neglected. The 25 acres we were going to plant had, over a period of years, produced only worthless scrub. Under the new regime waste on this scale would not be tolerated, and a contractor had cleared the area for the planting of a commercial crop.

Twenty-five acres is a fair piece of ground, especially when the lie of the land obscures the boundaries. Littered with the recently cut stumps of birches, and dotted with the ashes of fires where they had been burnt, the slopes of a small valley descended to a twisting burn.

Don cut the string on a bundle of Scots pine and laid the trees in a wooden box with a rope handle that he held out to me, together with a spade.

"Come wi' me, laddie, an' a'll show ye how it's din."

Following Don down the bank toward the burn, I could not help being amused by the spade. The blade was thin and extremely sharp, and had been reduced to half its original length by countless years of wear.

"Small spade, Don."

"Aye, laddie. When they get too smaa fer the nursery they're ideal fer plantin' wi'."

We all walked to the stream, except Will who remained at the top of the bank knocking in six foot high sticks where each of the first eight rows would end. About ten feet downhill of the first stick Will held a second stick upright. Tom, standing where the first row was to begin, directed Will to place the stick in line with the back one, before tapping it in with his spade. A few minor adjustments brought the sighting to Tom's satisfaction. He measured across to the next row.

Don took me to one side to show me the basic prin-ciples of planting a tree, while the rest of the team lined up behind their respective pairs of sticks and started planting up the hill. I had often wondered how conifers managed to grow in such straight lines – now I had the answer. Don stabbed his spade vertically into the ground twice in rapid succession, burying the short blade on each side of an L-shaped cut. The second cut made, Don pushed the handle away to his side and downward with a twist that lifted the top inches of ground like a tear in a piece of cloth. Taking a tree from his planting box he slid the roots under the spade and into the "tear", or notch as it is known. Holding the tree, Don withdrew the spade and firmed in the notch with a hefty tramp of his right heel.

A detailed explanation of the main points accompanied the demonstration: the tree must be upright, firmly planted at the correct depth, and the roots well buried. Straight rows and consistent spacings of about four feet six also mattered; weeds grew faster than trees, and when we were weeding in the summer any tree out of line or distance was much harder to find among the taller vegetation.

"Hae ye got that, laddie?"

Don's instructions had been clear and logical. All I had to do was plant a tree at the beginning of my row, pace four feet six forward, keeping in line with my sticks, and repeat the exercise until I reached the top of the hill.

"Sure, Don; no problem," I confidently replied. The job seemed simple.

"Fine! Awa ye go, laddie. Start yer row an' ah'll watch ye fer a wee while."

The words signalled the end of the lesson, and allowed me to get on with a job I was eager to tackle. The rest of the gang were scattered, roughly halfway up the hill ahead of me. I did not like being behind, but with the advantage of youth I would soon claw back their lead, overtake and arrive first at the top of the row. Grabbing my planting box I quickly lined up with my sticks. I applied considerable effort and concentration to the downward thrust of the lightweight spade to ensure that it would penetrate to its full depth in exactly the right place. If the sharp cutting edge had attained a speed of 50 mph as it hit the ground, it achieved 60 as it bounced back with a resounding whack. The instant reversal of direction had been caused by contact with a large stone hidden just beneath the surface. I winced as the shock of the impact transmitted through the handle, and up my arm to my shoulder.

"Och, laddie, ye've hitten a stane. Try a wee bit further back."

Left arm dangling numb and useless at my side, I worked back along the line, gingerly prodding the spade into the ground, only to find that the stone was in fact a huge rock that continued to the edge of the burn where it brazenly flaunted its might.

"Hell, it's a muckle yin, laddie. Try further for'ard."

I was beginning to think the rock had no end when the spade finally sank in, pushed by my foot to its depth. The numb ache in my arm dampened my competitive spirit, and I cut the notch without heroics. I soon learned that there was a knack to the clean opening of the notch – and that I did not possess it! However, a brief struggle later I had managed to plant my first tree – and it met with Don's approval.

Eyes on my sticks, I paced forward to my second tree's position. Some people complain bitterly that they have no luck – but I am not one of them. I have always had an abundance of it . . . all of it bad!

From what I could see around me the trees cleared for the planting had mainly been small, with the odd larger one dotted here and there. The largest tree in the whole of the area had been a beech some two feet in diameter – I knew that, because there in front of me was its stump, the centre of which coincided exactly with where my second tree . . .

I reached the top of the hill last, but wiser. The work could only get easier, and thankfully it did, as, throughout the day and row by row, I learned the art of the job.

The Physics Lesson

The "two stick" method of planting was simple and should have been foolproof, but it had one disadvantage: it took two to set up the sticks. Fine, as long as at least two finished together, but if only one finished someone had to leave their planting and walk back to the burn to sight in just one stick, and that proved a nuisance.

We had two laggards. Six of us planted as one, almost line abreast, but Johnny – The Professor – and Frank had to be different. The Professor lost time finicking over detail and would trail by one-third of the distance, while Frank struggled behind, repeatedly getting lost.

I could not understand Frank's difficulty. The white sticks stood out against their background, so all he had to do was keep in line with his pair. The Professor may have been fastidious, but Frank was not fussy at all. Given half a chance he would choose any two of the sixteen sticks as a pair and plant to them happily until stopped. For that reason the distance between the front and back sticks was kept to a minimum, but regular checking on Frank's movements was essential if he was to be contained in the right wood, never mind row.

Now in his late fifties, Frank's most prominent feature was his stomach, which exerted a constant testing pressure on the hip buttons of his bib and brace overalls. Small eyes peered from his round red face, which had an even ruddier round nose in the middle. Harmless and cheerful, Frank lacked an education, and became a handicap because of the attention he needed. The burden of overseeing Frank and pointing him in the right direction took its toll, and a degree of immunity to his plight was cultured. He was given assistance only when he asked.

Old Don apparently had his own special term of endearment for anyone that caused him excitement, and I was about to hear it for the first time: the word was "whore", but pronounced as a long "hoore".

The gang of six stood by the stream. With sticks set for the next rows, we relaxed for a mid-morning smoke and watched a small trout as it went about its business in the stream. Somewhere behind, out of sight and mind, Johnny and Frank were still planting.

"Och hell, ah dinnae believe it," moaned Don, who had made the mistake of looking back. "Look at yon auld fool."

Turning, we looked up the hill to see that Frank had veered onto a diagonal halfway up the bank. About 45 degrees off course, Frank had merrily planted across the gradient and into the other rows.

"Where in the hell are ye going, Frank!?" boomed Don.

"Wha's the matter like?"

"Ye're awa' tae hell. Ye've planted richt through oor last rows – ye stupid auld bugger!"

A sigh and a pause, and Frank grasped the peak of his cap, pushed it back, and vigorously scratched the top of his head. Carefully he stood his spade beside his last tree, screwed his cap on and stepped back at arm's length. To our great amusement he then squatted and sighted the sticks through the hole in the spade handle.

"Ah'm fair aback o' me stick!" retorted an indignant Frank.

"Ho! Ho! Hoo!" Don almost doubled up, roaring. "He! He! He! Hee! Ye could be onywhere in the bloody wood an' ye'd still be fair aback o' yer stick – ye donnert auld hoore ye . . . Ye've only got one stick in!"

Frank sighting his sticks

The day's fun did not end there; more was to follow.

The wind carried a chill prompting our retreat into the shelter of the scrub at lunch time. Finding a small clearing, we gathered around the remains of a fire that had last given off heat at least a year ago. Eating took priority over conversation, and only food passed the lips of the men that sat huddled around the ashes as though they still were hot.

Gentle Jake sat apart in silence, showing concern for his old tin flask; the weep had developed into a slow drip – it lived on borrowed time. However, the old tin flask did have one advantage over its modern replacement: it was nothing like as fragile. Our unknown predecessor who had built the fire had suffered the consequence of this failing. The evidence, in the form of a heap of shattered, silver glass fragments, lay in the ashes in front of me.

"Looks like you'll have to get a Thermos flask, Jake; they keep your tea hot much longer," I said, stating the obvious in an attempt to stimulate conversation. Jake flinched, but said nothing – it was Johnny, The Professor, who spoke.

"Know what it is that keeps the tea hot, lad?"

Suddenly I wished I had kept my big mouth shut; I had given The Professor the chance to test my scientific knowledge. Could I remember? My mind flashed back through time to school and the physics lab, where fortunately for me Mr Oliver was lecturing on that very subject.

"Yes," I replied, bursting to impress The Professor.

"What?"

"First, the vessel is made of glass, and glass is a poor conductor of heat. Secondly, the glass has a silver coating, and light, shiny surfaces reflect heat, whereas dark, dull ones absorb it. Thirdly, and most important, the vessel is of two-wall construction with a vacuum contained between the walls; heat cannot pass through a vacuum. It is a combination of these three basic rules of physics that keeps the contents hot."

My delivery had been eloquent, without the slightest sign of hesitation. But, curiously, The Professor had listened throughout with not a trace of expression on his face.

"You're wrong, son."

"Wrong? What do you mean wrong? – Course I'm not wrong!"

"You are!"

The Professor was taunting me – and stirring my ire. I knew what I had said was correct, but had I missed out some important detail? I did not think so, and glanced round the ring of faces hoping to find support... Nothing! Not a flicker. Puzzled, I conceded.

"All right! All right! So I'm wrong! Tell me what it is that keeps the tea hot."

I was allowed to simmer further as The Professor took a lengthy pause to enhance the dramatic effect.

"You know the glass inside the flask?"

"Yes."

"You know there's a little knob at the bottom?"

"Yes."

"That's what keeps your tea hot."

"The knob?"

"Yes. That little knob."

"For crying out loud! How the hell do you work that out?"

"It's simple – if it gets knocked off your tea goes cold!"

Aaaaaah! I screamed inwardly as everyone rolled off their butts laughing. Everyone, that is, except Frank, who did not seem to understand, Jake, who was worried about his leaking antique, and myself. There was a sort of logic in what had been said, but I did not find it amusing. It lit my fuse!

"No! No! No!" I cried in disbelief. "There's a gap between the two glass walls, and the knob is where the air is drawn out and the vacuum sealed in. If you accidentally knock off the knob you lose the vacuum – that's why the tea goes cold."

The Professor was noticeably embarrassed by my ignorance; he radiated disgust, and his eyes stared skyward flashing left and right. After what seemed like an age he fixed me with a glare of exasperation, rose with a sigh, and picked up a small piece of the broken glass from the pile in the ashes. Carefully holding the sliver between finger and thumb he went down on one knee, presenting the fragment before my eyes – and challenged: "Show me the gap between that."

Enough. Enough. I could not win and I knew it. I said nothing further. The others, minus Frank and Jake, had thoroughly enjoyed the entertainment, and expressed their appreciation by rolling around on the ground convulsing and hooting with laughter. In the end even The Professor managed to see the funny side, and joined in the merriment which he firmly believed was all at my expense.

Now I knew why the broad grin had split Will's face when he had said, "Knows about things scientific, does Johnny."

Chapter Three

The Contractor

I awakened early to greet a Friday – not just any Friday, but Friday 4 December 1959. The standard working week demanded forty-four hours spread over five and a half days. We were paid fortnightly. Strangely, the fortnight began on a Tuesday morning and ended the second following Monday night. We received our pay on the next Friday.

After working on Saturday morning, most of the team took the opportunity of going on through the afternoon for four hours overtime. I had done this and eagerly awaited my reward on this my first pay day. With luck on our side the planting would be finished tomorrow, and we would be going back "home" to continue thinning in the wood where Will and I had been peeling. Forestry was great and living up to all my expectations: the work was varied and satisfying, and the locations were continually changing. In less than three weeks I had learned a lot. On the practical side I could peel, lift and plant trees, and identify certain conifers. Most of all I had discovered more about the characters I worked with.

Will was of farming stock. The younger of two brothers,

he had been forced to work away from home as the farm was not large enough to support a father and two sons. I had learned he fell victim to peeler's wrist, a painful affliction that usually struck on his second day – explaining my turn on the peeler.

Frank had left his linesman's job with the County Council to work in the woods almost 20 years before. During his time on the estate he had selflessly denied himself a broad forestry education in favour of a singleminded specialist aim he had taken upon himself. His objective had been to achieve excellence in the art of disorientation, and his success – spectacular! Frank had yet another amusing quality: the ability to choose the wrong word of two that had a remotely similar pronunciation.

My guess about Ron proved correct. He was completely bald except for a closely cropped fringe below the back and sides of his cap. Ron gave the impression that money mattered by carefully rationing the contents of his cigarette packet. Ron's contribution to the profits of the tobacco company did not feature in the chairman's report – he existed on seven per day!

The world is not without injustice: Einstein was highly acclaimed and revered for his work in the field of physics, yet by comparison with The Professor he had merely tinkered with one of the sciences. Johnny's vast knowledge knew no bounds, and spread throughout the whole spectrum of science. He could, and regularly did, answer any question of natural or physical phenomena, yet he received the recognition of only a handful of men.

Extremely proud of his Scottish roots, Don often talked affectionately of the years he had spent north of the border. He had travelled widely in the course of his work, and built up a wealth of experience. Don had appointed himself as my working guardian at an early stage, and was only too pleased to pass on his considerable knowledge to his protégé. This I appreciated.

The most weathered face of all belonged to Tom, and yet, in his early thirties, he was the youngest of all. The exposed heights of his swarthy countenance were crazy-paved with a network of deep red capillaries. He stood slightly short of my five feet ten, but his rugged stocky frame was bulked by muscles developed with the demands of the work. Tom, like Don, had benefited from moving around with his work. He had a variety of skills and a sound knowledge of the job. He was the estate fencer, and undertook any repairs or renewals with the assistance of The Professor. I respected Tom: he helped me whenever he could, and paid attention to detail – like myself he smoked Woodbines.

The longest serving woodman on the estate was Jake. Soon to retire, he had left his native north Northumberland almost 25 years before, when he came to the estate as horseman.

It was a relaxed band of men that sat down for bait at lunch time. That Friday morning, the work had gone according to plan, making it certain that the planting would be finished the next day, after which the change of job and venue would be welcomed.

The demands that forestry made on the body's energy reserves created a healthy appetite, and sandwiches were attacked in an attempt to replenish the lost calories. Eating soon over, Tom rose, tossed me a Woodbine and said, "Come with me, Geoff. I'll take you to meet Joe."

Tom led the way into the scrub, which was quite dense in places. We planted trees at four and a half feet spacings, but nature did it differently. She broadcast seed en masse, and from there on it was up to the fittest to survive. The larger trees were not yet big enough to smother their smaller rivals, and our progress was slow. Tom ducked, weaved and barged his way through the profuse growth with his arms defensively crossed in front of his face. I followed a clear three paces behind to avoid the sting of the springy branches as they lashed back with venom. Thankfully the going suddenly improved as we came upon a well-used roe deer track. Conversation, as opposed to cursing, became possible and it transpired that Joe was the contractor who had cleared the area we were planting, and was a long-standing friend of Tom's.

Tom went on to explain contracting. Unlike ourselves, who were employed by the estate and paid an hourly rate, a contractor was a self-employed free agent who negotiated piecework rates with foresters on any estate. Work put out to contractors usually exceeded that which could be handled by the employed staff. Our work schedule bulged and we could not have been spared to clear even one of the 25 acres, which was only the first phase of a larger operation. Joe and his kind played a vital role, providing labour where and when necessary, helping foresters to complete their working plans.

Piecework sounded interesting. Joe had agreed a price per acre for the clearing, which simply meant that the harder he worked the more money he earned. I found the idea appealing; it seemed fair and logical that there should be a link between effort and pay.

Almost without warning we broke from the cover of scrub and found ourselves standing on the edge of a clearing, and there ahead was Joe, the man responsible for it. Joe did not believe in wasting time, and carried on with his work, unaware of any spectators. The flashing axe head had just demolished a 25 foot high birch with a few swings, and I watched as he tackled another. Angled downward, the first swing buried the blade almost halfway through the five inch butt. The second was horizontal and noticeably more vigorous, striking the birch close to ground level. A large chip of wood flew as the cutting edge sliced through to the bottom of the first cut. Using the axe left-handed, Joe repeated the actions on the other side of the tree; one cut more and the birch fell. I was watching an accomplished axeman. I was particularly amazed by Joe's ambidextrous ability: there had been no discernible difference when he had changed hands, and I could not decide whether he was naturally left- or right-handed. The axe swung with an impressive accuracy. Every cut counted, and was made with a precision that I knew could only be acquired with time.

With the butts held under his left arm Joe made for the nearest fire, where he axed the birches to manageable lengths and fed them to the flames.

"Not having any bait today, Joe?" Tom shouted, as we walked toward the fire.

"Hello, Tom. Why? What time is it?" Joe replied, turning with a huge smile on his face.

"It's gone quarter past twelve."

"Lost track of time; suppose I'd better have it."

I was introduced to Joe, who promptly grasped my hand and shook it as though I was a long lost friend; at least I assume that was his intention. The grip of his calloused hand was fierce, and my initial genuine smile must have waned to a sickly grin as I waited for bones to disintegrate. I swear I was standing only on the tips of both big toes by the time his hand relaxed and released mine. Joe turned and engaged Tom in animated conversation, the content of which escaped me as I discreetly assessed the damage to my hand behind my back.

It was December but Joe did not seem to have noticed. The almost sleeveless remains of a buttonless pale blue shirt hung in shreds from his shoulders. Any material that survived at the waist was tucked into a broad leather belt that supported a pair of khaki trousers in a similar condition. A long, frayed tear extended down the front of his right leg from above mid-thigh to mid-shin. Joe's apparel, or lack of it, revealed a large, strongly boned skeletal frame, powered by generous muscles that slid and knotted as he moved. Fine grey ash from the fire dulled his fair hair, and darkened every inch of exposed skin, especially that of his face, from which the whites of his eyes and his teeth flashed as he spoke. Beads of perspiration trickled across the furrows of his forehead and down his cheeks, cleaning the dust in their wake before dripping off his truculent, square jaw. Joe seemed oblivious of discomfort; he was a hard and tough, but friendly, man.

"Did you hear that Mary Hardwick had died, Tom?" Joe asked.

"Aye, someone told me, but ah cannot remember her."

"Ye must remember the Hardwicks, Tom; they used to farm Stonyleas. Mary was a very slender, religious woman, and her husband, Dicky, is so tightfisted that he makes a fish's arse look like a sieve."

"Yea, ah think ah know them, but only vaguely," replied Tom.

"Ye'll not have heard about her headstone, then?"

"Headstone? – No."

"Well, it's interesting . . . When Mary died, Dicky, true to form, scrounged the cheapest coffin he could find. And he reckoned that he could make another saving by getting old Josh Gibson to do the lettering on the headstone. Old Josh is a retired stonemason and a good hand with a chisel – so Dicky had a blank headstone delivered to old Josh's with a note of what he wanted engraved. Seeing as Mary devoted her life to the chapel, Dicky thought it would be fitting to have the words 'GOD SHE WAS THINE' inscribed at the top.

"Josh did his work, but the headstone arrived reading 'GOD SHE WAS THIN', which was perfectly true . . . but not quite what Dicky wanted. Having been diddled out of one letter, Dicky sent the headstone straight back to Josh with a message telling him that he'd missed off the 'E' . . . It came back engraved 'E GOD SHE WAS THIN'."

We were still laughing as we arrived at a fire that had died to a heap of embers. Beside it lay an old, much-abused shovel which had a long thin birch stem in place of the normal shaft. Joe stirred the embers with the shovel and, turning it over, gave it a smart knock on a stump. Using a hooked stick he then lowered a blackened kettle onto the prepared embers, and began rummaging in his haversack. I watched with increasing interest as his left hand produced a greaseproof paper parcel which yielded four large rashers of bacon. To my amazement Joe arranged the slices on the shovel, and placed it on the embers.

More secrets of the bait bag came to light as the bacon fried. A stained pint mug received an unmeasured quantity of tea leaves, shaken from a small glass jar. Next to appear was a large white loaf, which quickly lost four thick slices to a fearsome-looking sheath knife. Joe withdrew the bacon, turning the rashers with his knife, and I have to admit they did not look too bad as they lay on that filthy, battered shovel on the ground.

Fat from the frying collected in two pools on the shovel, either side of the midrib, and I looked on, mesmerised, as Joe's hand emerged from his haversack clutching two eggs . . . Never, I thought. Cracked on the edge of the shovel, the eggs were gently broken, one to each "dish", and the "frying pan" was returned to the heat. The hooked stick recovered the boiling kettle, and, protected by a rag

41

wrapped round his hand, Joe filled up his mug. There is an art to brewing tea, and the first suitable twig picked off the ground demonstrated the necessity of a thorough stir. A quick glance at the cooking revealed perfection. The "pan" was taken off the heat.

Joe frying his bacon and eggs

Joe divided the bacon between two slices of bread. Using a combination of fingers and knife he located an egg on each. Preparations, I thought, were completed with the topping of a second slice of bread, but I was wrong – or at least half wrong. Joe replaced one of the sandwiches on the shovel, and raised the handle so that the fat flowed forward around it to fry the bread. A forked stick pushed into the ground supported the shaft at the correct height after the shovel had been returned to the fire. Free of culinary chores for a moment Joe began to devour the first reward of his efforts with a lustiness a half-starved Labrador would have struggled to match.

Sadly it was time to leave as Joe's teeth hungrily crunched into his second course. Bidding him farewell we headed back to our work, and on the walk I had time to reflect on an experience that in my sixteen years of life I had seen nothing to compare with. It had taken only a few minutes – but what an education!

Piecework, as a system, explained Joe's eagerness to work to the extent that he lost track of time. It explained the sweat, the tattered clothing, and his attitude. He was paid by results, not by the hour. Time was irrelevant; he ate when convenient.

The shovel posed a problem: why did Joe have it? And there was the question of cleanliness, the importance of which had been instilled in me at an early age.

Tom supplied the answers when I voiced my doubts. Joe used the shovel to carry ashes from an old fire to start new ones as he worked through the scrub. With two or three shovelfuls of red-hot embers Joe had an instant fire. The long handle was a practical improvement; the heat from the fires was intense making the original short shaft impractical. Given this use the hygiene problem dissolved: there would be little chance of any organisms surviving the regular sterilisation process. I presume a similar fate would have befallen any bugs lurking on the stick that he plunged into the scalding tea.

Each day, it seemed, I learned not one, but many things. This was good, but it had a sobering effect on my youthful ego. I began to realise how little I actually knew. A glaring example of my ignorance had just confronted me. I had always looked upon the humble shovel as a hand tool of primeval invention, intended for an obvious purpose. The midrib, I had assumed, served only to strengthen, and

43

followed its accustomed shape so that the steel could accept the shaft. Given a lifetime I would never have guessed its real function: to keep eggs neatly separated while they fried. I was learning!

*　　*　　*

It was customary for the first person to notice Mr Shaw's approach to pass the word around. This was a simple safety precaution that ensured all hands were busily engaged when the boss arrived, and lessened the risk of complaint. Today was no exception in that respect; Tom saw him coming, and the warning had its effect; but there was a difference. The air of expectancy was there, but it lacked anxiety, in fact we were quite relaxed, and even I knew why. Jeff's arrival was welcomed because it was the one day per fortnight that he brought small, brown paper envelopes, and handed them out, one to each man. I was last, but accepted my first pay packet gratefully. I looked at the front. For two weeks work, plus eight hours overtime, I had earned the princely sum of eleven pounds, eleven shillings and six pence – before deductions, of course!

Chapter Four

The Sawmill

As it was winter, it was a safe bet that before long I would experience the sawmill. The rule was clear: "If it is raining in the morning, and looks like nothing else, go to the sawmill." That day had arrived – it was a Monday.

As I opened my curtains I saw raindrops firing from the heavens like bullets and exploding in their thousands on the tarmac outside the window, as if the clouds were on piecework. The weather had been remarkably subdued for the season, but today the truce had ended with a vengeance. A ghostly, blue-green glow from a street light pierced the gloom, allowing me to see the water as it ran off the tarmac on the road outside.

My mother and I had left our bungalow and moved to a semi-detached house. The distance was only three-quarters of a mile, but the difference was total. There had been no street lights or roads anywhere near the bungalow, and now we had neighbours on either side. Mother had decided that the bungalow was too large for just the two of us. It had been difficult for me to close the gate for the final time, and wrench myself away down the lane. Alas, I had no option.

*   *   *

I rode to work in the rain and was very uncomfortable by the time I turned onto the unmetalled road that cut through a wood of larches on its way to the sawmill. I had little in the way of protective clothing, and the rain quickly emphasised the need for proper motorcycle wear – I was soaked! Snaking my way between the puddles, I bumped along the road for a couple of hundred yards before coming to two wooden sheds in a clearing on my left. The nearest shed had gaps between the boards through which I could see stacks of sawn timber. The other smelled strongly of creosote. Behind the sheds and down a bank, tree trunks lay piled in a number of heaps. Lower still, at the bottom of the little valley, stood a large black Nissen hut – this had to be the sawmill. A stream flowed past it and disappeared amongst larches.

I was not alone for long; the rest of the crew soon arrived, and Will unlocked the sawmill door, uttering a few colourful words that expressed his thoughts on the morning. The hut was some 30 yards long and of semicircular section, clad with corrugated iron sheets supported on a framework of wood and steel.

Those who like myself had arrived on transport without a roof were suffering the consequences, and the first job was to light a log fire to get dried out. Following Don to the far end of the mill, we momentarily stopped beside the sawbench, and I was told that it had a twenty foot table and that the saw itself was a five-and-a-half foot Yankee. The information went over my head; I only remember being infused with a numbing apprehension as I looked at the vicious, chisel-like teeth . . . and thought about fingers.

Standing at the fire I willed the first flickers into action, but Frank, who had lit it, did something more positive. He applied used engine oil in liberal squirts from a large oil can. The fire was contained in an old brazier that stood on

46

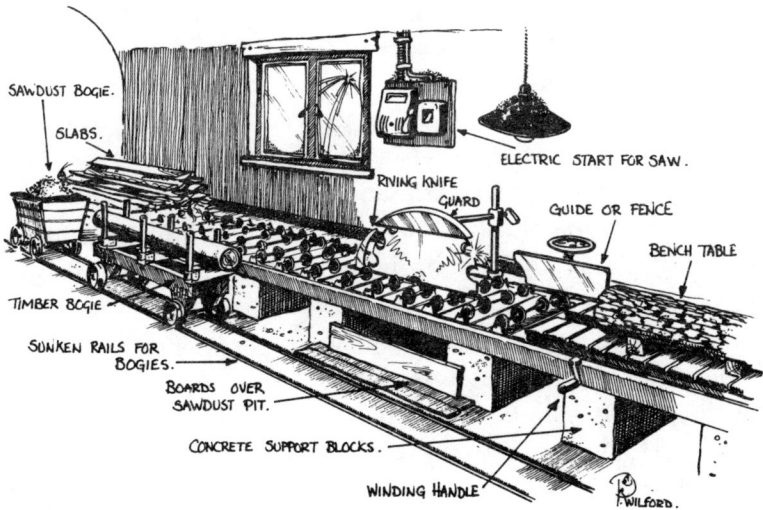

The sawmill

concrete at the far end of the mill, away from the door. The oil worked well, flames consuming it as they hungrily licked through the logs. Frank impatiently squirted more and more oil which helped the burning but, unfortunately, gave off a by-product: a thick, black, acrid smoke, that laboriously rose and collected under the roof – it had no option, there was no chimney.

Those of us who were wet were allowed to stay by the fire and dry out as the others prepared for the day's production. Don began by sharpening the saw. Will relieved Frank of the oil can and lubricated the bearings of the rollers that supported the table along its travel.

Ron pushed a scaled-down railway wagon to the saw along an equally miniature track that ran from the back of the mill, past the sawbench, to the door, and extended outside for 40 yards. Short, heavy planks removed from the floor revealed a pit beneath the saw where the sawdust collected. Descending into the pit Ron began to transfer the contents to the wagon using an outsize shovel. Watching as I dried, I could not help but wonder what Joe, with his ingenuity, could have cooked in a pan of those proportions.

Complying with the laws of science, the smoke displaced the heavier gases, like oxygen, before effecting a reluctant escape through the open door at the far end of the mill. Nine o'clock brought everyone to the fire for bait. Those who were more than a few feet away were guided there by the crackling of the logs – visibility was nil. The place was absolutely full of blinding, choking smoke. With posteriors lowered onto an array of logs, boxes and five gallon drums, we sat in a circle cocooned by the smog. The fire glowed, and the flames flickered in reflection on the haunted, bodyless faces that surrounded it in the gloom. Eyes wept and lungs coughed their protest, but nobody complained – it was accepted.

It was a relief to find I had to work with Tom and Johnny, the outside gang. The rain had eased and a little drizzle was preferable to the noxious alternative. Our job was to crosscut the trees in the heaps into lengths for conversion on the sawbench.

The job was more involved than I had first thought. Much had to be taken into consideration before a tree was marked into lengths. Species, quality and size all had bearing on the decision made by Tom. Once marked, the lengths were cut off with a crosscut saw and rolled to the edge of a loading dock which extended along the length of the outdoor railway lines. The lengths were loaded across skids onto a flat bogey, and pushed into the mill to be sawn. The crosscut saw was sharp, but the job of using it could not be described as fast. Cuts through the larger diameters took effort and patience, and whiled away the morning.

Lunch time bait in the mill was decidedly more pleasurable than the earlier experience. No more fuel had been added to the fire and it had been out long enough to allow the smoke to be replaced by beautifully fresh air. Seated as previously, we ate our bait in a much less hostile atmosphere. Although the torrential rain of the early morning had abated it had continued to threaten. The silence that usually accompanied the eating part of bait time was broken suddenly as phase two of the downpour was noisily unleashed on the roof – and by Don who commented:

"Bloody Hell! Will ye listen to that; it's fair pissin' doon like – "

Don was cut short in mid sentence; his words abruptly ended. Heads that turned questioning the unexpected silence found Don wearing a startled look, and frozen with a sandwich six inches away from his mouth. Guided by his open-mouthed gaze our eyes were cast to Jake, and thereon down to his bait bag . . . Eyes blinked in disbelief. It had to be an illusion. It could not be real – or could it? There, proudly sticking out of Jake's bait bag, was the unmistakable top of nothing less than . . . a Thermos flask! Jake had finally succumbed to one of the temptations of the twentieth century. Initially we were struck dumb, but Don recovered first, and blurted out:

"Fer crying out loud, Jake! What in the hell is that in yer bait bag?"

The query caused immediate embarrassment: Jake cringed at the words, hung his head and mumbled, "Doughter bought eet."

"Aboot time too! Ye'll like it, Jake; yer tea'll still be hot at lunch time," Don assured.

"Didn't want eet."

"Och why the hell no? Ye'd hae tae get yen soon: yer auld tin flask is knackered – it's fair pissin' yer tea oot like a Clydesdale stallion!"

Don took care to avoid understatement, and this last observation was an example, but in essence it was true. The old tin flask was finished. It stood on the floor in an ever-increasing pool of its own lifeblood – the tea that justified its very existence. Surely now, with his new acquisition, Jake would do the decent thing and retire his dear old friend to the dustbin.

Wrong again! Reaching forward, Jake clasped the flask between his gnarled old hands and drawing it toward himself he lovingly fondled it. Bent forward, with elbows resting on his thighs, he caressed the vessel and announced defiantly; "Nowt wrang that a spot o' solder wain't cure. Ah'll fettle eet th'neet."

Jake's distress was overt and painful; no further comment was voiced; the subject was closed.

It was time to resume work, but the rain did not care. It still lashed down, noisily venting its wrath on the corrugated sheeting that thwarted its intent. The consensus was that rain as heavy as this could not possibly last for long, and that we, the outside gang, should wait the shower out in the mill. I agreed on two counts. Firstly, I did not relish the idea of venturing outside and getting soaked once again; and secondly, I wanted to see the timber cut on the sawbench. Prior to now I had not been near a sawbench, and I was understandably keen to see the job being done. Another new experience . . . Little did I know!

The sawbench was simply constructed, and consisted of a wooden framework, some 40 feet in length, four in width and three in height, that supported metal rollers at regular intervals along its top. The wooden table was divided into two halves along its length, which were joined at the back by a single plank across its width. Running on the rollers, the split table carried the timber through the saw. This stationary circular steel plate was five feet six inches in diameter, and boasted wicked teeth.

I looked on as Don and Will rolled a length across the skids from the bogey to the bench. Eager as I was to see the action, I had an underlying fear of circular saws, because I knew of their reputation for claiming fingers. With Will at the front and Don at the rear the length of timber was carefully positioned on the table, and wedged to prevent it rolling. A nod from Don was the signal that Frank should start the saw, and that I should enter a desperate turmoil of internal conflict . . . Now I can laugh!

An electric motor powered the saw, and as Frank pressed down the handle of the switch on the wall the saw began to rotate with a deep hum. Slowly at first, the blade picked up speed, and the accompanying hum rose through the scales, and became louder and louder, crescendoing at a petrifying whine when running speed was reached. The acceleration was fast, but it lagged behind that of my heartbeat − I was frightened. I stood only feet away, and could sense every evil aspect of the saw's potential. The power, the speed, the teeth and the whine screamed out a warning that there would be no differentiation between wood, flesh or bone!

Timber met teeth as the table went forward, and there was a new noise as the saw ravenously cut into wood. There was also a change inside me − my fear left, swept aside by an inrushing flood of abject terror! I watched in horror as Will and Don pushed the log through the saw. They were so close − their hands on top of the length passing inches away from those teeth. The excessive stress stimulated my imagination and I could vividly see blood spurting, and fingers flying in all directions! Frozen by fear, my body responded, gushing adrenalin into the system: an essential defence mechanism that primes muscles for an explosion of activity. Unfortunately, in extreme cases, it can have an unwelcome side effect. Adrenalin does not discriminate; it motivates all muscles – even the ones over which we have no control! Instant in reaction to the stimulus were the involuntary muscles along the entire 22 feet of my bowels, and in the next few seconds I spent a harrowing eternity, desperately fighting to stop them expelling their contents . . . A battle I narrowly won!

I have no idea how many times Tom called my name, but the shout I heard was loud.

"Geoff! It's nearly stopped raining; we'd better get on!"

The words were wonderful: they jerked me from my trance and allowed me to move once again – and move I did! I needed no second telling. Never in my life have I been so pleased to vacate shelter, and work in beautiful drizzle!

Back outside, I could still hear the sound of the saw as it ripped its way through timber, but mercifully I was freed of the vision of the gore and severed digits. Involved in my work, my fears subsided, and by three o'clock my terror had diluted to anxiety. I was reassured by the fact that, as yet, nobody had been rushed to hospital, but I still harboured nagging worries that were hard to suppress. Trying to conceal my true feelings with a disguise of nonchalance, I commented, "That saw in the mill is a hell of a size, Tom."

"Five-and-a-half feet – that's normal."

"Rips through the timber at a heck of a rate."

"Faster than this," Tom replied, nodding to the crosscut we were alternately pulling.

"Do Will and Don have to work with their hands so near the saw?" I asked, instantly regretting my phraseology – I had blown it!

"First time you've been near a sawbench when it's working, Geoff?"

"Yes. I haven't even seen one like that before," came my reply.

Tom's face broke into a broad grin. He stopped pulling on the handle of the saw, looked me in the eye, and said:

"Fair scares the shit out of you! – Doesn't it?"

The Sawmill: Again

I opened the curtains to yesterday . . . It may have been the next morning, but the view through the window was identical to that of 24 hours ago: rain – solid water. The sawmill again.

Memories of the previous day flooded back, and I wondered what ordeals this new day held in store. More toxic gases from the fire? Those dreaded teeth again? Or would it be some other form of torture?

Only slightly wet in places I arrived at the mill. The rain had almost ceased by the time I left home and had held off for the duration of the journey before re-starting. This proved doubly beneficial, because as nobody was wet the fire was not necessary, and we sat down to bait in air tainted only by the smell of freshly sawn timber – a great improvement on yesterday.

"Shit! The buggor!"

Heads whipped round in the direction of the un-expected expletives to find Jake holding a parcel of saturated sandwiches that dripped liquid onto the floor . . . Not a word was uttered, not a breath taken or a movement made.

"The sod – an' ah fettled eet last neet!"

We all knew that the old tin flask was beyond repair, but his love for it, coupled with a stubborn streak, had driven Jake beyond the bounds of possibility. Now he was paying for being unrealistic.

Dropping the sodden packet back into his bait bag, Jake sighed as he grasped the handle of the offending vessel. With a grunt, he rose, and strode slowly and deliberately to the far end of the mill. Looking on spellbound, we exchanged rapid glances of amazement as Jake carefully sited his old companion in the centre of a large chopping block.

I am sure the thoughts running through my mind were common to all of us: I could not believe my eyes! Jake was never going to . . . He could not . . . Not gentle Jake . . . Especially not to his beloved . . .

Eyes bulged and chins dropped as Jake reached forward, took hold of the handle of a seven-pound felling axe, and drew it toward himself. Axe head between his feet; the shaft on his thigh; Jake addressed his intended victim with a long, cold stare. It was out of this world. Fantasy. It could not happen. Yet there we were, rendered incapable, watching it all unfold. With eyes locked onto his target Jake spat on each hand before grasping the shaft of the axe. Drawing himself to his full height he prepared to deliver the weapon with an exacting determination that I have not seen equalled. The backward element of the swing was relatively slow and deliberate; the axe head leaving his feet and gradually accelerating until it apexed high in the air where an unintelligible curse imparted an astonishing velocity to the downward stroke. The old tin flask gasped a dull thwack and spurted what blood it had left high in the air as it died beneath the back of an axe that connected squarely with its top! The strike was lethal. Jake, not satisfied, carried on, and the swing, curse, spurt cycle was repeated and repeated with ever-increasing speed, ferocity and excitement, until the hail of blows had reduced the veteran vessel to an object without height.

Dropping the axe, Jake stooped, picked up the disc of metal and walked out through the door where he launched

The end of the old flask

it into oblivion with a farewell string of obscenities!

The events had been totally out of character – Jake was a mild man. However, he had a breaking point, like any other human being, and the infidelity of his trusted friend had been too much to bear.

Muttering to himself as he walked back, Jake smacked his hands together with satisfaction before heavily taking his seat.

"Buggor wain't dee that ag'in!" he asserted, gratefully reaching for the saviour of his day – the vacuum flask he had so persistently refused.

The demise of Jake's antique had been spectacular, enthralling – and a gross miscarriage of justice . . . It was the two day old Thermos that was broken! The old tin flask had been innocent.

The torture I had expected early in the day now began,

and as with the smoke we all suffered together. Poor old Jake was stupefied with anguish at the thought of what he had done: forehead bowed into one hand, his other held the shattered Thermos which he shook and rattled in disbelief.

Jake was undoubtedly in pain, but it was nothing compared to what the rest of us were experiencing. Having just sat through the most hilarious drama of our lives, without drawing as much as a breath, we were heavily pregnant with mirth. The silence was electric. Faces reddened and contorted; lips tightened and twitched; eyes threatened to pop from their sockets, and cheeks strained with the effort of containing the laughter. If Jake had not possessed a sense of humour we would all have died. Fortunately, after a severely testing age, his face betrayed a slight grin which was more than enough to trigger the explosion of our pressurised hoots as we fell off our seats, and roared and howled our approval.

Some performance! Absolutely unforgettable!

Chapter Five

Thinning

If patience is a virtue, woodland owners have to be the most virtuous men on earth! Preparation for planting – clearing, ditching and fencing – the planting itself and subsequent maintenance all cost money, and it can be two decades before the first thinning produces a financial return – that requires patience. The proceeds from that one operation may, or may not, cover the cost of the work involved, but at least it is a step in the right direction. Forestry is not for short-sighted investors: the only fast bucks a woodland owner ever sees have four legs, testicles and an appetite for trees!

A forester aims to produce an even, final crop of large trees with straight, gradually tapering stems, standing at around 80 to 120 to the acre. At the time I started working it was considered best to plant at spacings of four feet six inches square, a density of over two thousand trees per acre, and reduce the numbers by phased, selective operations known as thinning.

Close planting encourages rapid increase in height, and therefore little taper on the trunk; it also restricts branch growth, resulting in smaller knots in the timber. It does, however, create the need for man's attention, as natural selection would allow the most vigorous and not necessarily the best trees to dominate.

On the estate the forester controlled the selection process with his ultimate goal in mind, and as a bonus, the operation produced an early (on forestry's timescale) financial return from the trees removed. Demand for the small timber that thinning produced was high; some became fencing material, but the majority disappeared deep beneath the ground into the mines. First thinning usually took place 15 to 20 years after planting, when approximately a third of the crop would be removed. This exercise was repeated as required over the years until the final crop was left to grow on before clear felling.

Trees to be removed in the thinning were marked by slicing off a length of bark on two sides of the tree with a slash knife, or similar tool. Marking was considered a skilled job, and was usually carried out by the forester himself, a foreman or trusted employee. The principle involved the selection of trees of good potential and the marking of any inferior tree that threatened competition; this ensured that the better trees would grow unimpeded. In addition dead and sick trees, and those likely to succumb to suppression, would also be marked.

Around me the white blazes revealed by the marking stood out on the trees under sentence – at last my day had arrived. Forestry, from the moment it was considered as an occupation, meant only one thing to me – timber felling. I was well aware of the many and varied operations in the production of timber, from the sowing of seed in the nursery to finished articles leaving the sawmill, but my mind painted only one picture: that of trees falling. There lay the challenge – and the excitement.

Every wood on the estate was identified by name, and

the wood where I had spent my first working day helping Will to peel was known as The Major's. Thinning could be carried out at any time of the year, and we were back there to continue the work which had been left unfinished because of the planting.

The Major's cut a long, steeply sided dene below the side of a byroad it followed for about a mile. A small burn ran its length, blindly following gravity's meandering path through patches of level ground where the hillsides had been eroded. This once-frightening mixture of six trees was being thinned for the third time, and some of them, especially those growing in the lower areas, had been drawn up to a height of 60 feet. I was understandably keen to see my first one fall.

"Which way are we going to fell this yen, Frank?" asked Don, pointing the handle of his axe to a Norway spruce of one foot diameter.

Frank gave the spruce a calculating scan from bottom to top and down again before walking to the butt, where he deposited the crosscut saw and a folded sack he had been carrying. Don gave me a mischievous wink as Frank entered into what I can only describe as a rite of decision making. Taking position with his back firmly against the trunk, Frank pushed back the peak of his cap, and took stock of the situation in a 90 degree segment to his front. Responsibility and Frank were not natural partners, and the burden showed as he weighed up the possibilities. The deliberations were assisted by a stimulating scratch of his dishevelled grey thatch and accompanied by long sighs of "Aye", "Uh-huh" and "Mmmm" before he shuffled round the tree to the next stage where the procedure was repeated. The final "Mmmm" at the end of stage four was quickly followed by a tightening tug on the peak of his cap which signified that all relevant details had been assimilated and acted upon: the decision had been made!

"Where are we going to drop it then, Frank?" Don asked impishly.

"There's ownly one bloody way it'll gan, Don!"

"Where's that, Frank?"

"Eithor ower there, or ower there!" insisted Frank, indicating a totally different direction with each arm.

"Ho! Ho! Hoo!" chortled Don gleefully. "You remember that, laddie. When ye've no choice – ye hev the pick o' twae!"

Don had instigated the amusement entirely for my benefit; he did not need to involve Frank; it was simply a manifestation of a youthful devilment he had retained, and could not resist.

Still chuckling away to himself, Don spat on his hands, grasped his axe and laid into the root buttresses, or toes, with a succession of vertical and horizontal swings. The sharp edge sliced deeply and precisely. Large chips of white wood flew until the base of the tree was almost perfectly round, left standing in contrast to the wildly irregular section of its root. Don, aware of his attentive onlooker, stood back and explained.

"That's her rounded up, laddie. Ye'll notice yer bottom cuts should slope wi' the surrounding ground; now all a've tae dae is put in the gob. The gob is important; it controls the direction of fall, an' ye must get it richt. It's a 'V'-shaped notch we cut across the butt at 90 degrees to where we want the tree to drop; it works like a hinge. It should be as low as possible, an' go about a fifth of the way into the tree. Watch an' ah'll show ye."

Returning to the work in hand, Don cast a glance along the intended line of fall before slicing out a clean notch with a few economical swings of his axe. Axe-work complete, Don and Frank knelt on folded sacks to each side of the tree with their backs to the gob. Crosscut saw between them, they entered the teeth, and with reciprocating pulls on the handles began to cut toward the notch. Progress through the cut was rapid, and, as the saw drew worms of white wood with every stroke, I began to understand why conifers are commonly referred to as softwoods.

The suspense increased as the saw neared the gob. Eyes glued to the top of the tree, I watched intently for the slightest sign of movement – Yes! There it was! It moved – I was sure that the top moved an inch, but Don and Frank kept sawing – There it went again. It was definitely going!

"She's awa', Frank," Don said in an acquired matter-of-fact tone, and the pair rose and stood back. Don's non-chalance was understandable – he had seen it all before thousands of times, but for me it was an exciting first time . . .

There was a strange silence; time appeared to stand still. The top eased a fraction, stopped, and eased again, almost but not quite coming to a halt. Slowly and gracefully the tree began to leave the vertical, hesitating slightly before gaining speed with every submissive crack and groan as the uncut timber behind the notch yielded. Once committed to the downward plunge the acceleration increased; only the air sighed its resistance through the branches, pushing them and the slender top back and upward . . . A muffled thud brought to an end a spectacle I will never forget, but all was not ideal: I had watched, not made it happen.

I was privileged in that I had been allowed to stand and watch the whole felling operation, and it was not until it was over that I realised that I was the only idle body. The others were busily rounding up, and laying gobs into marked trees in readiness for Don and Frank with the crosscut.

"How's that fer ye, laddie?" beamed Don.

"Great," I replied, dying to get to work with the brand new four-and-a-half pound axe I had been given.

"Ye've tae think aboot it afore ye drop yen, laddie; ye cannae drop them onywhere. Every yen has tae be winched up the bank to the road, an' if they're felled awkward it can make the extraction difficult."

I could see how a lack of thought, or care, when felling a tree could prove expensive later, and, looking at the axe in my hand, I was wondering how long it would be before that important decision would be mine.

Don read my thoughts and continued, "Dinnae fret, laddie, ye'll soon learn tae use yer axe. First ye've tae learn tae sned; Tom'll show ye how."

The pair moved off to the next tree as Tom approached.

"Can I have a look at your axe, Geoff?" Tom asked, his arm outstretched.

"Sure," I replied, offering him my prized new possession.

Tom scrutinised the edge of the axe, frowned, and pulled a flat round sharpening stone from his pocket.

"Hell, Geoff, it's as thick as a bull's lug! But new uns usually are – it'll take time."

Tom spat on the stone, and as he started honing the edge with a vigorous circular action I began to realise the significance and versatility of saliva. On the stone it acted as a lubricant, yet applied to the hands it served the opposite purpose, and I could not help but ponder the implications for forestry if the supply ever dried up!

"No, it's no good," sighed Tom as he checked the results of his efforts. "Ah'll give you the stone, and every bait time you'll have to work at it 'till it's sharp."

The only experience I had had of an axe was a small, single-handed one we used at home to chop kindling, and by comparison my new axe was very sharp indeed.

"Is it really that bad, Tom?" I queried, impatient to get

on with the job that I could see everyone else doing. They were dressing out or snedding: cutting off the branches of the trees that had fallen to Don, Frank and the crosscut.

"It'll do for a start, but once you get the hang of the job you'll need it like this," Tom assured, offering his axe for examination.

"Heck it's like a razor!" I gasped, not believing an axe could be so sharp.

"Course it is, Geoff. It needs to be; it makes life easier. Nobody in their right mind works with a blunt axe."

Tom had been emphasising his words by using his axe to cleanly shave the hairs off the back of his arm, and he went on to detail the requirements of a good axe and show me how to achieve them. The main problem with my brand new Elwell was not so much that the edge was blunt, but that the angle was too thick. As it was it would have a tendency to cut out, and away from the tree when snedding, leaving snags, rather than a neat cut, flush with the trunk.

"Oh yes, and while we're on the subject of axes, Geoff, I'd better point out something, just in case you might make a mistake. Whatever else you may do, Geoff – never ever use an axe that's not your own! Don't even pick one up – the penalty can be death!"

The fact that I failed to grasp the seriousness of the crime became manifest in the puzzled frown my face adopted.

"Do you remember when we were in the nursery, Geoff?"

"Yes."

"And when we were planting?"

"Yes."

"Did you notice that everyone had their own spades?"

"Yes, I suppose they did," I agreed after a pause.

"Yes, Geoff. Everyone has their own tools. The spades

63

you have used are yours for as long as you work here, like that new axe, but a spade is one thing and an axe is something totally different: it is the most personal tool a woodman uses!"

I could comprehend some of what Tom was trying to say – but the death penalty? Tom patiently pressed on.

"Geoff. How long is it going to take you to get your axe like mine?"

"Umm . . . err . . . a while," I replied vaguely.

"Yes, Geoff, a while – a bloody long while! It could take months; even if you spend every spare moment working on it. Imagine the day when, after countless hours, you finally have an axe to be proud of, the edge like a razor, and the angle exact." Tom paused momentarily. "What would you do if someone picked it up, swung at a toe, and hit a bloody big stone that took a chunk out of the middle of the edge?"

"I think I'd kill them, Tom!"

"Right! You've got the message, Geoff. Here, I'll show you how to sned."

Moving to the branches nearest the butt of the spruce Tom began to dress out. He held the axe with his hands well up the shaft, and swiftly severed the branches with quick, short strokes. No effort was wasted: each stroke had only enough energy to carry it through the cut, made flush with the wood of the trunk. The branches grew in distinct whorls and were cut off, usually singly, in a sequence with a lively, fluent action, the axe head never striking twice in the same plane. A running commentary, explaining the technique, accompanied the brief but thorough demonstration.

"Practice is the best teacher, Geoff. I'll leave you to it. Just remember the basics, and you'll soon pick it up."

The instruction illustrated a sequence in three parts:

firstly, the uppermost branches, then downward between the tree and yourself, and finally the far side. This was purely for ease of working, the methodical way to do the job. Another aspect was one of safety: when snedding on the nearside, the axe must be past your legs before it strikes the branch. Failure to observe this simple rule of common sense could result in a painful reminder if the axe bounced off and made contact with a leg. Tom had assured me that most woodmen bore such scars as testimony to a momentary lapse. The main demand of snedding, however, was the ability to wield the axe with accuracy, consistent throughout the continually changing angles of stroke as the tree was stripped of its branches.

Tom walked off to get on with his own work, leaving me to continue on the spruce. He left me highly impressed with the demonstration of his artistry with the most important tool of his chosen trade. He made it look so easy, but it did not fool me – I was learning! In my short working past I had seen "easy" jobs – planting had looked easy until I had tried it. Joe, Don and now Tom made axework look simple, but there lay the difference: they only made it look easy because of their highly developed skills. Now, lower in gullibility, I addressed the work in hand free of youthful delusions. My caution proved justified, and I was neither disappointed nor surprised when my first stroke landed only roughly in the area of the targetted branch, which to my amazement fell off, leaving a snag. The snag lost some of its length to the second stroke, and the third finished the job – three strokes, all for one branch – practice was needed!

I pressed on, determined to acquire at least some of the technique before I reached the top of the spruce, which lay feet away in distance, but looked hours away in the future. Having to think about my every move slowed my rate, and it took an extravagance of strokes and a generous helping of time before I reached my destination – the point where the trunk tapered to two inches. Here the snedding ended, and one casual swing of the axe removed the top – I knew that: I had seen it done many times by those working around me by the time my turn came. Taking deliberate aim at a point on the top I unleashed the axe . . . I missed by a miserable two inches – and my second shot hit the same place! . . . Still the top survived – a severe test of patience lay ahead!

Tree by tree my accuracy, technique and speed improved, and by lunch time I had made progress; slight it may have been, but nevertheless, progress it was.

"How are ye getting on, laddie?" inquired Don, as he pared slices off his Condor Bar and rubbed them out in his hand.

"Better than when I first started, thanks," I replied, studiously honing my axe as Tom had instructed.

"Och, laddie, dinnae worry. It takes time, but soon ye'll be as guid as onybody," promised Don.

Tom reached forward, relieved me of my axe, and held it toward Don who sat contentedly drawing the flame from a match deep into the tobacco in his pipe.

"Have you seen the lad's axe, Don?"

"Nae, ah havnae," Don replied, accepting the offer. Holding the axe at face level, Don cast his expert eye along that all-important edge. An educated right thumb further tested the tool and brought the comment: "Och hell laddie! Ye'll need tae be very carefy tae watch yer legs. It'll no cut ye, but the bruise'll be an awfy sight! He, he, he, hee! Never mind, laddie, a'll show ye what ye'll be able to do after a wee while."

Don rose, returned my blunt instrument with a smile of encouragement, and walked with his axe to a freshly cut stump. Bending forward, he stabbed his axe lightly into the stump, making three separate cuts a few inches apart. I watched, puzzled, as Don carefully inserted a matchstick perpendicularly into each cut. Satisfied with the arrangement Don stood back and, grasping his pipe like a gear lever, crashed it into the extreme left corner of his mouth. A habitual spit on each hand completed the formalities, and solved my mystery . . . He was going to split . . . with an axe . . . never! Axe held across his front Don eyed the target for a second or two before one, two, three rapid swings, loosed without pause, rained down on the stump.

"There ye are, laddie; ye'll need tae practise 'till ye can do that."

I had remained seated at some distance, from where it was difficult to see the extent of Don's accuracy. Two of the matchsticks had moved; he had been close, but just how close I could not tell. Submitting to curiosity I got up and walked toward the stump, where I sank to my haunches. What I saw surprised me – Don had not been close – he had been absolutely spot on! I had difficulty believing my eyes as I looked at the two matches I had seen move – both were split in two, virtually centrally along their length, and the top half of the third was minus a sliver. The precision was remarkable, especially with a tool like an axe, where the head was controlled at such speed at a full arm and shaft's length from the body. Of one thing I was sure: it was no fluke. Don was no idiot; he would not have gone to the trouble of setting up the matches to make a fool of himself. No, this was the norm, the standard I must aim for.

My thoughts were cut short by the resumption of activity, as the afternoon's work got under way. Don and Frank continued with the felling, leaving the rest of us to dress out and heap the trees in readiness for extraction. After dressing out, the heavier trees were rolled onto a short length of timber so that their butts were clear of the ground, and the smaller trees carried, or otherwise man-handled, into heaps around them.

The afternoon was pleasantly cool, ideal working conditions for the physical exertion demanded by the job, and I began to enjoy myself. My new axe gradually relinquished the total control it had over me and accepted a compromise of mutual understanding – I was going to win!

This was forestry at its best; trees were falling, and for once it was a physical spectacle and not a mental vision. Felling formed the first link in the chain of harvesting, an operation that had a special, poignant significance. The gathering of any crop offered the ultimate reward: the bounty nature paid out for man's efforts. The fact that this thinning was only an interim crop did not detract in the slightest – far from it, the trees we were cutting had been planted some 40 years ago! It takes a special breed! It must have been a forester who coined the phrase: "Expectation is greater than realisation."

Time sped by, as it has the uncanny knack of doing when you are enjoying yourself, and we all pressed on with a sense of urgency into deteriorating light. The daylight may have been relenting, but Don did not: he carried on to the finish with his remorseless abuse of Frank.

"Which way are we goin' tae fell this yen, Frank?"

The answer never varied.

"There's ownly one bloody way it'll gan – eithor . . ."

I felt sorry for poor old Frank, and began to wonder if Don could take stick as well as hand it out. Enlightenment lurked in the non-too-distant future.

Sod's Law

Cold air snatched at my lungs as I stepped out into the December morning. I found the blue-green light from the street lamps alien and was unused to hearing the noise of a diesel engine as a bus passed the gate and growled its way up the hill. It should have been pitch-black and quiet – that I was accustomed to. Back home at the bungalow I had fortunately been insulated by distance from both sources of what had now infiltrated my life as forms of pollution.

I may have been used to the black of the long winter nights, but I detested them, and had done so from an early age. The short days spoilt my fun. September marked the beginning of the decline; often a pleasant and enjoyable month, it nevertheless heralded winter's ominous approach. The light, eroding from both ends of the day, increasingly curtailed my activities until the only hours of free time coinciding with daylight were at the weekend – I loathed it and felt thoroughly cheated. A love-hate relationship developed with the winter solstice. When distant, I dreaded its approach, but as it neared I could not wait for it to arrive, pass and disappear, obliterated by its own murk.

The Bantam was not exactly inspired either – it lacked spark. Not the spark of vitality – alas it never had that – but the one that fired the mixture in the cylinder and powered the piston down. It made no attempt to start – not a flicker. Experience had taught me that a bump start would not suffice; the bank that the bus had climbed began at the gate, and I was not tempted to push the bike against the force of gravity in a futile effort to start it on the downhill run. A brisk clean with a wire brush and the plug was replaced.

Repeated digs at the kick-start spun the engine over, and soft words of encouragement rose to threatening curses before a spluttered response greeted my ears. Frantic choke and throttle adjustment kept the piston moving, coughing a blue fog out of the exhaust, until the firing became normal. Pulling in the clutch, I engaged bottom gear, and accelerated out through the gateway toward the sweeping left-hand bend at the nearby road junction.

With the benefit of hindsight it could be argued that the bike's behaviour had not been unreasonable, that it had been trying to tell me something.

I was aware of the sharpness of the morning but, unlike the bike, I found it stimulating and was raring to go. The knowledge I lacked might have seemed trivial: it was merely that there had been a shower through the night, and that the temperature had dropped low enough to convert the water to a solid state. These minor details were filled in as I leaned the bike into the corner and applied some power to a rear wheel on ice.

It all happened very quickly. The rubber found no grip, and deposited both machine and rider gracefully on the tarmac. Fortunately our speed had not been high, but the momentum was sufficient to propel the pair of us in an exhilarating spin across the rink. I felt no pain, only the discomfort of embarrassment, which quickly subsided when I realised there were no spectators. In a way this was a pity, because on technical merit we may have been lacking, but the artistic impression would have earned a 6.0 from the Russian judge! Some start to the day.

The annoyance at my failure to notice the conditions was cut short by music from the Bantam's exhaust – the engine was still running. Declutching, I lifted the bike, threw a leg over and eased away from the scene on a new-laid carpet of caution. Over the bridge and out of the village I was in a different county, and world for that matter. Slowly, with feet trailing off the bike, I climbed into Northumberland, leaving the eerie haze of the artificial light hanging in the valley.

The ride had been anything but pleasant, and I rounded the final bend with a sense of relief, immediately dashed by the sight of the Land Rover parked ahead on the roadside . . . the boss was there! Jeff was not always on the job first thing in the morning, but today was that special occasion he just had to be there for. He had to be there to welcome me on the first time I was to arrive late – but with good excuse.

The heads that turned briefly in my direction as I pulled up belonged to bodies in attitude of prayer. The small congregation stood in ritual salute, encircling the object of their worship. Fingers spread on the downturned palms of extended hands shielded faces from the radiated powers of their idol – a fire – a beautiful, life-restoring fire. The journey had left me frozen stiff, but not entirely without movement – the lower jaw still worked making my teeth chatter in an uncontrollable frenzy. Any thoughts of admonition for my tardiness were dispelled from the numbed grey matter. All efforts, mental and physical, were concentrated on transporting the body to the source of the rejuvenating heat. At least that was my plan – Jeff had other ideas. He had noted my arrival, looked at his watch and was making toward me with determined strides . . . I awaited a bollocking.

"Morning, son, enjoy your ride? Grand day for it."

"N-not a l-lot," I shivered in staccato reply through clashing teeth.

"Don't stop the bike. Follow me, you're cutting Christmas trees today."

Jeff's greeting had been cheerful, and mentally I thawed,

71

warmed by the lack of remonstration as I followed the Land Rover that transported Jeff and Ron in enviable comfort.

Adorned with seasonal embellishment, the tall trees in The Major's slipped past, and we continued down the road until a hard ride cut down steeply to our left marking the end of the wood. Turning down, we pulled up on an old stone bridge that spanned the stream in the lower reaches of its ramble through the trees. Craftsmen had built the bridge; skill, pride and time had been mixed into the mortar, that was obvious, yet strangely it seemed wasted – a work of art hidden in the middle of a wood.

"There they are, son," Jeff smiled, nodding downstream as he handed me a Bushman saw.

The scene my eyes beheld was picturesque, steeped in atmosphere yet clinically appropriate for my first day cutting Christmas trees. It was a picture with qualities no artist could be expected to capture. Downstream of the bridge, standing branch to branch, an abundance of young Norway spruce clad the lower ground. Popular because of its neat and profuse branching habit, the Norway had usurped the indigenous Scots pine as the Christmas tree of the moment. But I looked upon no ordinary Christmas trees: the ones before me were greatly enhanced – pre-decorated with cotton wool and tinsel.

The night-time shower had fallen as rain at home, but over the higher elevations of Northumberland it had been a different story. It had begun with the same steady rain, shed from the trees a drop to each needle, but later the altitude took effect, causing the precipitation to change and fall as white, blanketing flakes of frozen vapour. The shower had lingered a while before moving on silently into the night, clearing the sky behind it. With the cloud cover gone, the temperature had dropped rapidly, catching and freezing water droplets in suspension.

Leaving us with our instructions, Jeff climbed into the Land Rover, engaged four-wheel drive and scrambled up the bank only just ahead of a chasing spray of snow from the spinning wheels. We had to cut three hundred trees between four and six feet in height. Only the best were to be chosen, as this was a special order – no problem: the trees in front of us were certainly special!

"Bloody Christmas trees," growled Ron, scowling as he drew unusually heavily on his Player's. "It's always like this when you're cutting the bloody things. It's Sod's law."

The words filled me with alarm. Ron was not noted for verbosity; his words were sparse, and his vocabulary lacked the rich colour of normal forestry jargon. Never once had I heard him swear – things must be bad!

After a hurried bait, we exchanged a cursory shrug of the shoulders and stepped forward to the fence between us and the trees. Climbing the rails was like negotiating the border of a Christmas card and entering the abstract of the scene beyond. The feeling, however, was short-lived. The reality returned and the beauty decreased with each protesting crunch as the crust on the three inches of snow yielded beneath the weight of advancing feet. Still frozen, I welcomed the prospect of bodily activity, which could only encourage the circulation out of sluggishness and generate heat. Confronting the trees, we agreed on a strategy: I would work from the stream to mid-distance, and Ron would tackle the other half.

The trees had been planted about three feet apart; the idea being that over a number of years every other row, and every other tree in the remaining rows, could be cropped as Christmas trees. In this way 75 per cent could be removed to bring an early return, and still leave a crop to grow on for timber . . . or so the theory read. Nature penned the difficulties into the small print on the practical side of the contract. Trees are all individuals, and as such are allowed to exercise their right to vary in size and form – a right to complicate. If all the trees had been between four and six feet high, and if all had been of uniform quality, then I would have been working elsewhere . . . luck like that and I had not been introduced. There was no alternative; the only way we were going to cut trees to specification was by penetrating their icy defences and selecting them.

The first candidate that drew my attention stood in plaintive humility, dwarfed by extrovert neighbours. Sorry as I felt for myself some of my heart went out to that unassuming little tree as I neared with saw in hand. It just stood there in innocence and tranquillity – unaware of its fate. The conditions did not favour kneeling, and I went down on my haunches with the saw at the ready. I became aware of a developing feeling of guilt as I swept the lower branches upward with my left arm, and pushed the saw toward the infant trunk. The guilt vanished as the tree fought back. Some of the branches sprang out from under my arm, catapulting a vicious shower of snow and ice at my head. There was nothing general about the direction; the volley followed a precise path, and most of it found the target – the gap between my collar and neck. I did three things instantaneously: I swore, I drew in my head, and I hunched my shoulders. This proved highly successful in trapping the missiles against the bare skin of my neck – they were heat-seeking, and desperately in need of it. Hell, they were cold!

Automatically I knelt, head-down into the snow, relaxing only when I was satisfied that gravity would help with the evacuation of the foreign bodies. Index fingers together at the back of my neck, I carefully drew them forward around my collar, and removed the cause of discomfort.

Relief flowed through my body in a warming glow, but it did not reach my knees for they were cold, even colder than when I had arrived. Looking down I discovered why. Preoccupied with my predicament, my presence of mind must have lapsed when I rolled forward – a stupid mistake to make. My knees had pierced the covering crust of snow, and were immersed in icy water into which dissolved any remaining trace of sympathy I had for that tree. It was now fair game. Guarding against a repeat attack, I carefully put my hand below the branches, pushed up high, and grasped tightly to prevent their escape. This worked well: not a single piece of ice or snow went anywhere near my neck . . . They went up my sleeve instead! I could not win.

The tree lost its battle, as did the next and the next, and as I pressed on, cutting and dragging out, I began to understand what it was about the job that made Ron swear. It mattered not in the slightest how you pushed your way among the trees: your vulnerable parts were always targeted. Whether you went forwards, backwards, sideways, crouched or erect, they always got you in the neck or up a sleeve.

Lunch time brought a well-earned respite, and found a miserable pair clasping flask tops in frozen hands. They were all we had that gave off heat.

"Surely it's not always like this, Ron?" I queried.

"Too true it is, lad! Sod's law dictates that it is," Ron moaned, killing further dialogue beneath his stamping feet.

My hands thawed, warmed by the flask-top of tea, but from the knees down I remained frozen. Feet, stuck on the outer ends of limbs, could be expected to suffer from cold, but my knees, having undergone a water-cooling process, had also lost their sense of being. I had to do something. Finishing my tea, I joined Ron in the foot-stamping exercise, but there was no improvement. Youthful resilience appeared to have deserted me, and a strangeness prevailed, which eventually prompted investigation.

Learning came by degrees. I had first discovered the trees' overt methods of retaliation, and had concentrated my defence against them, but I had failed to notice the clandestine activities of their resistance movement. Every round of frozen ammunition that missed my neck or sleeve could be found in my wellingtons – packed solidly around my legs. Some day this was turning out to be! The boots, impossible to remove, were in turn lifted behind and hooked over a rail in the fence, whereupon a pounding fist on the leg evicted the nuisance. The day could only improve.

I was right; the improvement materialised as we marched back to re-engage the enemy. A deathly still had preoccupied the morning, and we were caught, totally by surprise, as the warm fresh breath of a westerly breeze caressed our backs. Together we stopped, exchanged brief glances of astonishment, and turned to look back as if expecting to see a physical explanation of the sudden change. Of course, there was none. The air had sprung from nowhere; it was one of those rare occasions when there was an abrupt, and unexpected change in the day . . . I drank to that change. My body temperature had consistently fallen since I had left home. Now bordering on hypothermia I relaxed and opened myself to the warmth. It was most welcome. At least I thought so.

"Sod's bloody law, lad; I told you! This job is all about Sod's law," groaned Ron in obvious displeasure. "It's bloody typical!"

The words left me speechless. Two more swear words; provoked by an unexpected bonus? It did not make sense.

I restarted work with a lightened heart, but with an unenlightened brain; I could not understand why a change for the better should induce a curse. One of the disparities Ron and I shared was experience – he had it and I did not.

The difference between an experienced and an inexperienced person was simply, in this case, that the former was aware of the ramifications of Sod's law, whereas the latter was about to be educated! Single-mindedness had blinkered me to the overall effect of a sudden jump in temperature. I had overlooked the fact that the warmth was not exclusive to us. The festive dressing festooning the trees quickly began to soften, and it did not take long for the message to seep through . . . literally. I was as quick on the uptake as my clothing was of its intake of every drop of water as it was made available – thawed by that breeze! Understanding dawned. It was a race against time, and the longer the job took, the more the ice and snow would melt. Right from the start the going was against us, and hopelessly outclassed by the opposition we finished a poor second, saturated, and heartily sick in defeat.

The job finished, I gratefully turned my back on those trees, but trudging away wearily in wellingtons sloshing with water I had to concede that the day did have one good point: I had not been punished for arriving late, as Jeff had said nothing.

Sod's law had not cropped up in any science lesson at school, and when Ron first mentioned it I was caught in ignorance. I had not heard of it. That was earlier. Since then, I had had time to reflect on the events of the day, and decide that the extensive exposure I had suffered to the wide variety of applications qualified me as an authority on the subject. Sod had opened the door for me that morning and influenced my every move, everything was attributable to him – he must have enjoyed the day.

Strange that the first thing to prise a mild expletive from Ron should be the humble little Christmas tree. As for me, I simply hoped never to see one of the ****** things, again – not ever!!!

Chapter Six

The Outfit

Gone was the winter solstice. Gone was the scourge of Christmas trees. Spring was in the offing, the outlook brightening.

We had slithered with time through winter's ice, snow and slush. The old year had drifted onto a page of history, taking Jake with it into well-earned retirement – long might he enjoy it!

Steadily the days had whittled at the nights as summer approached, and, no longer curtailed by a lack of light, our working day had increased to "proper" time. We now started at 7.30 a.m., but there was consolation. The ten-minute breaks became official and the lunch time half-hour grew to 40 minutes – but it was never referred to as such – it was always "A half an hour and ten minutes."

We had moved on. After the felling in The Major's all the trees had been extracted by winch up the bank onto the level ground by the roadside where conversion had taken place.

A full-length, dressed-out tree was not a readily saleable product; the markets were more exacting, requiring conversion. All the thinnings cut in the time I had worked were softwoods, the majority of which were channelled, according to species and size, into two main outlets. The mines had an insatiable appetite for timber, and devoured a healthy and varied diet by the ton. Round timber was taken in the form of pit-props in an assortment of sizes, invariably peeled, and sawn timber was consumed as splits and crown-trees. A six foot length of timber with a five to six inch top diameter, peeled and sawn in half along its length, produced two splits; and crown-trees, also six feet long, were square sawn to five by two-and-a-half inches.

Fencing created a considerable demand for timber, from small round stakes only requiring pointing on a sawbench, to square sawn posts and rails in a variety of sizes.

Conversion of thinnings was usually carried out on site using basic equipment: a push-bench and peeler. A push-bench was simply a small sawbench on which the timber was pushed across a fixed table, and – like the peeler – the push-bench was belt-driven by the tractor. Normal procedure had been that Don and Will, as sawyer and tractor-man, enlisted a third gang member, and beginning at one end of the wood with the push-bench, they cross-cut and converted round timber into sawn as they worked their way to the other. Pit-props were sorted and stacked to be peeled in a separate operation.

That is how it had been in The Major's where I had been the third man, but the system, fraught with inherent inefficiencies, had been modernised by a touch of ingenuity. One of the main problems was that the tractor was tied up over the time it took to make the three working journeys through the wood. First with the push-bench, cross-cutting and sawing; then with the peeler; and finally sawing again when the splits, which had to be peeled first, were sawn down the middle on the bench. Wasted man-power had not escaped notice either. A three-man gang was one too many once the cross-cutting was finished, and peeling was a single-handed job turned into a two-man operation by the threat of accidents. But hopefully those problems had been left behind with the sawdust and shavings in The Major's.

We had moved across the road and had thinned Narrow Wood, an appropriately named belt of smaller conifers that ran down the roadside opposite The Major's. The extraction had been completed and the trees lay in heaps ready for conversion, but this time it was to be different – we had the outfit.

The solution to the problems had overtly begged discovery; it was so obvious, it had been overlooked. Both belt-driven, the push-bench and peeler were each mounted on wooden skids, which allowed them to be dragged through the wood behind the tractor, but until now the logic of mounting them in tandem on one long pair of skids had evaded discovery. Two heavy skids, sixteen feet in length, had been cut from a selected larch butt in the sawmill, and the push-bench, stripped of its own skids, had been bolted firmly at one end. After being tightened against a belt placed over the two pulleys, the peeler was similarly secured at the opposite end. A long belt from the tractor over the push-bench pulley now drove both machines, and theoretically the conversion operations could be carried out in one pass by a gang of only three men – that would be progress! The modifications had cost a few bolts, a little time and some of our own timber, a small price to pay for the potential benefits, but it had yet to be put to the test.

That moment had arrived. The outfit, staked to the ground to prevent it moving to the pull of the tractor belt, waited in readiness alongside the butts of the trees in the first heap. The saw and peeler, rotating in unison, invited the challenge.

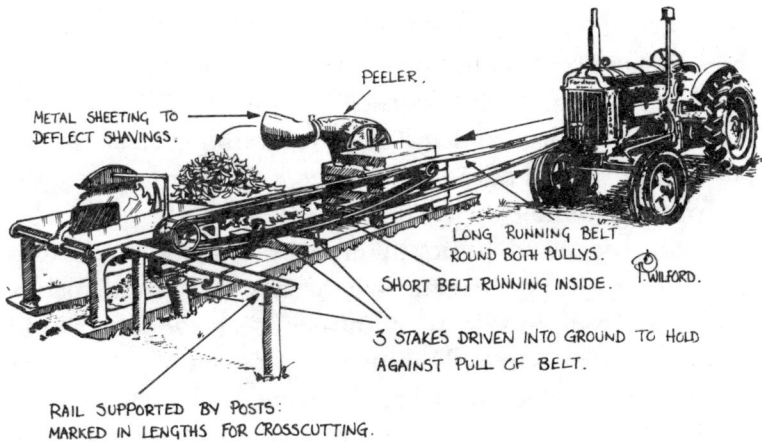

Labels on the illustration:
- PEELER.
- METAL SHEETING TO DEFLECT SHAVINGS.
- LONG RUNNING BELT ROUND BOTH PULLYS.
- SHORT BELT RUNNING INSIDE.
- T.WILFORD.
- 3 STAKES DRIVEN INTO GROUND TO HOLD AGAINST PULL OF BELT.
- RAIL SUPPORTED BY POSTS: MARKED IN LENGTHS FOR CROSSCUTTING.

The outfit

With Don at the butt, Will a few feet further up and myself at the tail end, the first tree was lifted across the bench. Using only his "eye" and experience, Don made an instant assessment of what the first length would produce and measured it off against the relevant mark on a piece of rail supported at the side of the bench. A quick push forward cut off the length which Don laid beside the bench while Will and I pushed the tree over the table for the next cut. Work on the push-bench was surprisingly swift; the saw was kept extremely sharp, Don made sure of it, and the heap of trees soon dwindled. The advent of the outfit had changed nothing in this element of the job, and it was never intended to. The timber to be sawn was still stacked within easy reach of the bench, and the props tossed aside. The impact the outfit was to make only came after the cross-cutting was finished.

During the cross-cutting, the peeler, coupled to the bench, had been humming in idle rotation, but now, instead of stacking the props and splits to be peeled in a later pass, I could get on with the job there and then, at the same time as Don and Will were sawing – and all with one tractor. The props and splits had been tossed beside the peeler and all I had to do was peel everything, stacking the props once instead of the usual twice, and return the lengths for splitting to the bench – that was progress.

Not a day passed without comment extolling the virtues of the outfit. The increase in efficiency was incalculable. Each time we moved on to another heap we left behind finished articles. The only return to any stack of timber was for loading. In all a huge improvement.

My mettle had been severely tested by forestry's ultimate deterrent – Christmas trees, but I had survived unscathed, and enjoyed every minute of conversion. I derived a great satisfaction from it, gained in experience and met new faces.

Popular at the time with the local farmers were split larch posts; indeed we had difficulty meeting the demand. If ever we managed to produce a small stock invariably a tractor and trailer would pull in and relieve us of every one. Larch was favoured because of its well-deserved reputation for durability without being creosoted. Unfortunately this quality only applied to the European larch. The Japanese larch, and a cross between the two, the hybrid larch, were no better than any other softwood. However, the farmers' ears were deaf to advice: a larch was a larch to them and so we gave them what they asked for.

Split posts measured five or five-and-a-half feet in length, and were quickly sawn from round timber on the push-bench. A length of three to four inch top diameter would make one post after having a thin slab removed, a four to five inch top ripped in half produced two, and a six inch top yielded four when quartered. They sold at two shillings each, and were in great demand.

My education had been intensive and without break

since I had started work, but I made no complaint; I was there to learn. The subject did not stop at forestry. The boundaries of education widened to encompass the curriculum of country life in general, and it was the farmers calling in for their posts that taught me many lessons.

Farmers, or their hands, were regular visitors, calling in at the sawmill not only for posts and rails, but for other one-off orders. They frequently came into the woods for split posts, and if time was spare when they were passing they would often stop just to have a chat – it seemed important. The welcome they received would have befitted a long-lost friend, and after the initial trading of salutations the conversation usually became more serious as the local news was updated.

I had not known any of the farmers in that area, so each face was new, as were their names. The faces sported the expected variation of features, characteristically weather-beaten, and there was nothing unusual about them. The same could not be said of their names: Watson Morrison, Rogerson Bates, Taylor Henderson, Forster Miller and – by contrast – John Brown. I could understand John Brown, a simple Christian name preceding a surname, but the others . . . ?

Puzzled, I sought clarification. "Will, why do the farmers round here have such strange names?"

"Strange?"

"Yes, strange! They have a surname for a Christian name."

"Oh, I see what you mean, lad. But that's not uncommon around here; it often happens.

"But why?" I persisted.

"Ah suppose it's a way of preservin' a family name, lad. It usually happens when the first son is given his Mother's maiden name as his Christian name, but it's not always like that. It may be some other family name, and it's not always the first son that's given it."

I juggled names in my mind. The same ruling applied in my case, and I would have answered to Scott Surtees, which would have been reasonable; but imagine being the first male issue of a union between a Miss Featherstonehaugh and a Mr Sattersthwaite! That would have been unfair. Endless combinations of names flitted through my brain in the course of the day – and I saved them for future relief from the boredom of the nursery.

Chapter Seven

The Country Life

"Listen – Listen!" Head turned to one side, Tom fired the words with an urgency that brought the bait time banter to an abrupt halt. Everything stopped as ears strained to pick up sound – sure enough it was there, distant but unmistakable: the sound of foxhounds running.

"Hounds!" Will shouted as we scrambled to our feet. We had all heard the faint strains, and followed Will in a hasty retreat to the road at the edge of the wood.

It was a Saturday lunch time and we were having a break from thinning in Old Ben Hardy's, a small, low-lying wood stuck near the bottom end of Narrow Wood like an oversized dot on an exclamation mark. Nothing could be seen from the roadside, as we were too low down, and could only listen to the distant music. The hounds were in full cry, according to opinion, somewhere in the region of the North Drive – and heading our way. The excitement increased.

"They must be over the road and in the top end of Narrow Wood," Tom guessed.

"Sounds like it," Will agreed. "And look – there's Charlie!"

The fox, commonly referred to as Charlie in country circles, had run the length of Narrow Wood, and we watched as he slipped out at the bottom corner, crossed the road and disappeared into the compartment of infamous Christmas trees in The Major's. Running well in pressing pursuit, the hounds spilled out over the road and fell silent, casting in all directions. Left on their own the hounds worked well for a few minutes before the huntsman galloped down the roadside verge and joined them.

"Did you see him?" he asked.

"Yes," we chorused.

"Where did he go?"

"Right through the middle of where the hounds are casting," Will replied.

Strangely, that was the end of that particular hunt. The hounds had covered a fair distance running well on a good scent, but try as they might, they could make nothing more of it. The scent had mysteriously vanished, and Charles with it.

Shaking his head, the huntsman reluctantly accepted defeat, collected his hounds and moved off to draw for another fox. All of the followers spoke to us as the field passed, provoking a light-hearted exchange of comments with some of the local farmers as to the worth of the hounds.

"Where the hell is he going now? There's nowt along there!" Will stressed, bemused, as the pack and field turned off the road through a gate at the end of Old Ben's.

It was indeed a strange sight to see a pack of foxhounds, in search of a fox, turn its back on acres of fox-harbouring woodland and head for open agricultural land – well, almost open. Will's observation carried a lot of truth, but lacked strict accuracy – something was there . . .

A track hugged a dry stone wall as it made its way in a straight line along the low side of two fields, giving access to a farmstead some 200 yards distant. Snuggled up to it was a covert of postage-stamp proportions. It was nothing out of the ordinary for trees to be planted near a steading because they afforded shelter, but not this patch – it was too small, and defied justification. Each of the four containing walls of the covert could have been no more than 30 yards in length.

"He's never going to . . ." Will faltered as the pack neared the farmstead.

"Lieu in – try!" encouraged the huntsman, signalling the direction to the pack with an underarm sweep of his hand. Fourteen couple of hounds poured over the wall and into the tiny covert – and at the same time, four foxes left – one by each corner! The clamour was immediate as hounds gave tongue to their bewilderment at the confusion of red-hot scents, and split in three directions, ignoring the fox that vacated the haven by the corner nearest to us.

"I don't bloody believe it!" Will exclaimed, speaking for us all as the huntsman faced the decision of which fox to chase.

Four foxes were three too many for the hunt. Each one had demonstrated a healthy instinct for self-preservation by leaving the covert with haste, our fox included – the one running the farm track. He began his flight with the same extended strides as the others, but as backward glances revealed no pursuers, his urgency was diminished, and his pace soon slowed to a lope. The whippers-in struggled with their work to stop and collect the divided elements of the pack, and return them to the huntsman who had made his choice: the fox heading our way. We stood aside from the gate.

We held our breath as a leisurely walk brought the fox

to the gate. Head throught the bars, he stopped and flashed the amber of his eyes in our direction. Satisfied that we posed no threat, his body and brush followed his mask through the bars, and with a second casual glance at his audience he turned, sat and looked through the gate. He could not have been more than 12 feet away, but he ignored our presence, directing what little interest he appeared to muster along the track toward the steading. Full pack together, the huntsman lost no further time in putting hounds onto the track where they spoke without conviction to a patchy scent.

The fox

"Aye, they're away again," Will commented, "but the scent'll be poor on the road. And look at yon cheeky bugger – he knows it."

Will, of course, was referring to the fox at the gate,

which had steadfastly maintained an air of indifference with his rear firmly fixed to the ground – until hounds gave tongue to his line. This new development spurred Charlie into an instant flurry of activity. He drew in his front legs as much as an inch; he raised his head slightly, and for the first time condescended to half-heartedly spectate the local event of the day. The fact that he was to take the leading part in the action did not overawe him. He just sat there on a cushion of apathy, watching the hounds working his line, with a barely discernible increase in interest as they neared. Nobody said a word; we stood stock-still, wondering how long the fox would be content to sit and watch his pursuers gain ground. Some 200 yards had separated fox and hounds, but the hounds' perseverance on the poor scent steadily reduced the distance by 50 and then another 50 yards – with not the slightest effect on Charlie.

It was only when the distance between the two parties had been halved that Charlie began to pay any real attention. He pricked his ears, cocked his head and appeared thoroughly to enjoy the entertainment as the pack relentlessly closed in. A 50 yards' start was all Charlie seemed to need, and he only moved when that point was reached. Grudgingly, he heaved his body onto all four legs, and turned to stand facing us. He appeared to give a shrug of resignation and I will swear that he winked before a flick of his brush sent him over the road at a trot. Across the road, the gate offered an open invitation into the Christmas trees, the sanctuary he entered, never to be seen again.

Once in the wood, Charles II was spirited away in exactly the same manner as his predecessor had been. The scent either vanished, or had never existed. It was eerie. A mysterious agent lurked in those trees – they had the ability to chill in more ways than one.

I had heard it said that any quarry pursued by the

overwhelming odds of a pack of hounds fled in blind panic fuelled by abject terror, but there was little evidence to substantiate that theory from what we had seen. Far from it. A cooler, more calculating character than Charlie would have been hard to find. He knew no panic – but suddenly I did.

We had been engrossed. The mysterious events had promoted lively discussion and we were oblivious to all else. Time was forgotten.

I had no chance to give a warning. When I first saw the Land Rover it was slithering to a halt on four locked wheels. Time, camouflaged by diversion, had slipped past without anyone noticing. I dared not look at my watch. It had to be at least a quarter past one, and there we were, naked of excuse, standing outside the wood we should have been working in for the past three-quarters of an hour. My heart sank; we would all be sacked.

The whole of my short working life flashed through my mind in a fraction of a second. In reverse chronological order the replay skipped back through time to the interviews, and I heard those words again. I felt physically sick. Shuddering at the thought of the future, I salvaged small consolation from my plight – at least now the dialogue would differ.

"Good morning, Mr Forester. I'm Geoff Surtees. I've called in . . ."

"Yes, lad we do, but we're looking . . ."

"But I am experienced. I can plant, I can peel, I can . . . I can . . ."

"Aye lad, maybe ye can. And a'll tell ye what else ye can do – Bugger off. Ye were ownly there three months, and ye got the bloody sack."

I knew that word travelled fast, and to avoid total humiliation I was going to have to move further afield, ahead of the telegraph forwarding my disgrace, perhaps as far as Australia – after all, passages were assisted.

The door burst open before the Land Rover actually stopped, and the seat behind the steering wheel ejected a serious-faced Jeff, who immediately tackled Will.

"What's going on?"

Accepting fate, I banished the cowed attitude my mind and body had assumed. I would stand, proud and erect, and take my punishment. At least Will was going to get it first.

" 'Going on?' ye ask," Will retorted fearlessly. "Wish to hell I knew. Makes no sense; ye won't believe it."

A glancing blow to my head from a single spruce needle would have floored me as Will began to give Jeff a belated commentary on the hunting in lucid detail.

"Hell! I wish I'd seen that, and four in there is unbeliev-able, unless they were playing dominoes," laughed Jeff as he took his seat in the Land Rover. "Never mind. You may get another hunt later."

The hunt had hacked off up the road and Jeff, about to disappear in the opposite direction, engaged first gear and departed.

I was amazed. I had not been sacked, and neither had anyone else. There was no invective, no punishment, no mention of work or extended bait times.

My brain had finally reached saturation point with infor-mation I had gleaned over a period, that would not piece together.

I did not understand the bias of general conversation toward parochial matters: I was puzzled by the strangely close relationship between everyone involved with the estate and its environs, regardless of rank or capacity. A local farmer acquiring a new implement, or a woodman moving from one estate to another, caused more vibration of the vocal cords than issues like a general election. I needed some answers.

The days passed, and subtle questions dropped into conversations at opportune moments proved rewarding. Each forthcoming ray of enlightenment pieced together into a hazy picture, that gradually cleared.

It was a morning about a week after the hunt that I awoke to the dawn. Not to the rising of the sun – that happened later, but to the dawning of an understanding of true country life.

Forestry was like any other job in one respect: at the end of the day we all went home. However, I, like those in the vast majority of occupations, left my workplace to go home; my workmates did not. There lay the difference. Not only were they employed by the estate, they also lived there, retiring in the evenings to the isolation of their often small, tied cottages that had only recently been searched out by the creeping copper tendrils of the national grid. Socialising outside the estate was not easy; the weekend's excursion to the local pub, three rural miles away, required organisation. Any interested party without motorised transport had to make their way to an agreed point on a circuitous route taken by the evening bus – Jeff's Land Rover.

The estate was not only their work; it was their home, their life and their world, and little of consequence happened outside it. A happy community flourished; a kindred spirit bonded those who devoted their lives to the land. The disciples of forestry were gregarious, everybody knew everybody, and news travelled fast, keeping track of the movements of distant brethren.

Most landowners were wealthy, but rarely with the income from woodland. For many, forestry was little better than a self-financing hobby, and this was reflected in their struggle to pay a penny more than the minimum basic wage. Despite this, landowners often assumed a voluntary patriarchal role, showing a genuine concern for the wellbeing of their *family*, and spread tables with food and drink for all the estate staff to celebrate special occasions.

My fears of instant dismissal for confusing priorities between work and fox-hunting had no foundation whatsoever. In fact I would have been courting reprimand had I been at work. Strange and confusing, until the last piece of the jigsaw fell into place.

The interest shown in a hunt was not confined to the actual hunting of the fox. That was of secondary importance. The prime motivating factor was more obscure: not so much what was happening, but where. The hounds were welcome visitors, and when they came it became an occasion, an event in the calendar of estate life, and we were expected to show an interest. Exactly the same reasoning applied to shooting days, when invitations to beat were expected to be both tendered and accepted.

At last I had the answers. I had gained an understanding of real country life – the fraternity of true countrymen, most of whom, born within the system, had known, and would know, little else. They were happy, and enjoyed simple lives, free from the pressures of aspiration. They possessed little except contentment, but that they owned in abundance.

Chapter Eight

Just a Little Too Far

Special was the morning. Fresh, alert and eager to partake of the day, I walked out into the embrace of spring. It had been flexing its muscles for some time – at long last it had made its move. Pleasant days of unseasonally dry weather had followed the chill and mist of the early winter mornings, but this morning was different. Warmth and brightness cheered as spring finally toppled the shaky authority of winter on the wane. The longer days were nearing. It was great to be alive.

High on elation, I opened the bike's throttle and powered into Northumberland with an exhilarating new confidence and respect for my motorbike. The air recognised it, whistling its apologies for grasping at my clothing as it was hopelessly tossed aside. The gradients acknowledged it, the uphill levelling out and shortening, while the downhill steepened and lengthened. Bends straightened, miles shrank, and time took on a new scale. I felt on top of the world, devouring the miles, thoroughly enjoying every corner and straight of the ride, and, not surprisingly, arriving early at work. A flick of a switch turned the ignition off, bringing to an end one of the most pleasurable rides of my life. I turned the petrol tap off, and took great care to rest the bike safely against the wall – I had no intention of scratching my gleaming new Matchless 250.

Careful nursing had failed the Bantam, whose condition had steadily deteriorated to the point where its unreliability could no longer be tolerated. Faced with the decision of either spending considerable money on repairs, or retiring the Bantam in favour of new machinery, I had chosen the latter option.

Today we were working in High House Gill, a small valley that began placidly with gently sloping sides, but changed character rapidly with every angular turn as it scarred an ever-deepening gouge through rolling pasture land. A crow could glide an effortless mile from top to bottom of the valley, but for the small trout in its stream the journey was doubled through waters tumbling an erratic path.

When the season's planting had been completed, we had followed on with the pugilistic-sounding job of beating up. The Gill had been clear felled of its mature crop three years previously, and had subsequently been planted. Beating up involved nothing more violent than walking the rows and replacing the trees that had failed to take, or had suffered irreparable damage by vermin.

My beating up debut had been made acres ago, and on the many miles I had walked infilling the "misses" it had become apparent that the forester has few furry friends. Vital to any conifer is the central leading shoot and bud, essential to maintain its distinctly monopodial, or single, straight-stemmed habit. Vermin attack in this area results in stunted and deformed growth, usually rendering the tree worthless to the forester. Damage varies in severity and form, and one soon learns to identify the culprit by its *modus operandum*.

Forestry's arch enemy was once the rabbit, mainly because of sheer weight of numbers, but by the time I had started working at the estate myxomatosis had reduced its population to a level where a sighting was a talking point – it was a pest of the past. At least the once-dreaded rabbit preferred grass and other light vegetation, and usually it took a good covering of snow to turn its attentions to trees for food. When prolonged bad weather dictated, the rabbit would nibble at bark, but if greenery was in reach it was content to browse on the newest shoots, eating its way round the tree without showing particular preference for the terminal shoot. But not so the hare: it craved destruction.

It seemed that the hare knew of the vacuum created by the disappearance of the rabbit, and had decided to fill it. Its numbers were increasing at an alarming rate, and so too was the damage to trees. The hare was nothing better than a despicable vandal that delighted in wanton destruction, and worse, it had the audacity to advertise the fact. There was no mistaking the marks of the ravager: pointless waste and a methodical approach were its hallmarks.

Only one part of a tree interested the hare, and that it neatly severed with razor-sharp incisors. It was, of course, the vital terminal shoot on which so much depended. The less-important laterals were rarely touched. The obsessive preference of the hare for that one and only upright shoot I found difficult to understand. Could it taste better? Un-likely, I thought, because if it did why did the hare not eat it? There lay the malice: the hare did not eat it – at least not more than one bite per tree. It simply snipped off the shoot tops, which fell to the ground like the business card of a tradesman, leaving a clean oblique cut on the tree. Anyone could be excused for mistaking hare damage for the work of some idle person casually swiping off the tops with a sharp pocket knife, because at first sight that is exactly what it looks like. However, you can see that this is not the case when the cut-off top is held in place against the tree top: the two bits do not match, as the jagged cut on the top does not fit the clean cut on the tree. I have never caught a hare in the act, but I will give it the benefit of the doubt that it actually eats the one severing bite, before embarking on a havoc-wreaking mission along the row, repeating its evil deeds with undwindling dedication. The hare may be classified as game, but to the forester it is vermin of the lowest order, and deservedly so.

Another pest is the field vole. Seldom seen as it scurries about its high-speed life under the cover of fallen vegetation, it balances its diet with helpings of basal bark from young trees. Its taste for this delicacy kills many trees, either by ring barking (eating a ring around the trunk), or by creating an open wound through which a fungal parasite can enter. On grassy sites in particular, a close inspection of a dead but apparently undamaged tree often reveals the tell-tale gnawings of the vole. Small the creature may be, insignificant it is not.

When we were planting new trees there was a minimum daily target of 600 trees per man; any less and questions were asked. Beating up was more relaxed and pleasantly different. The most satisfying reward offered was the walking of a row, or two together, without replacing a single "miss".

Planted three years ago, the hybrid larches at the top of the Gill had taken well and grown vigorously, suffering only slight damage by vermin. This was the second beating up, and I had no complaints as I walked between the rows, checking the trees to my right and left; my legs were working more than my spade, and that was how it should be.

Experience had taught me that forestry was not the most straightforward of occupations, and I should not have been surprised by what brought me to a sudden halt. The damage that confronted me was blatant; no close inspection was needed. I looked with some despair at a patch of eight of the over-four-feet-high larches, all of which were in need of replacement. Not one had fallen victim of the rabbit, hare or vole: they had succumbed to the forester's largest foe, the roe deer, and it was the male of the species that had been responsible. Both sexes are harmful, browsing on trees, but the buck has the irksome habit of inflicting further ruinous injury, defining the boundaries of his territory. I stood in the middle of a bad example. Commonly the buck uses only one or two trees in an area, but here, for reasons best known to the culprit, that was not enough.

Many males in the animal kingdom zealously defend their territory, warning potential intruders against trespass by scent-marking at strategic points along the boundaries, but it is the roe's method of depositing his warning that causes this problem. Sophisticated, the roe does not deign to use urine or excrement: he has a scent gland specifically for the purpose. Unfortunately for the tree and forester, the gland is located between the small antlers of the roe, and the only way he can leave his odorous message is by contact with some object. Closely set, the antlers narrow down the choice of suitable warning-bearers.

Pushing firmly against a young tree held between his antlers, the roe stimulates the gland and deposits his scent with an energetic rubbing action which frays off the bark in shreds. The endeavours of the roe to protect his patch mount as he travels his boundaries. The rough antlers demolish tree after tree until the buck is satisfied that his message will be noticed. The roe is graceful and fleet of foot, and the cost of fencing against its agility is often dismissed as prohibitive. Their numbers need to be strictly controlled.

Our roe appeared to be quite content with the one thing that came readily to head in the wood. I did not dwell, either mentally or physically, in the area of damage, but left that behind as I worked on until lunch time when I joined Will, Don and Frank in taking a seat against the boles of the semi-mature larches of Sawmill Wood which shared a boundary with the Gill.

Although it had been a beautiful morning, there was trouble on the horizon: a heavy black cloud hung with a menace over the lower end of the Gill. When we reached there, the beating up would be finished, and our reward would be a solid, six-week stint in the nursery. I banished the thought, and concentrated on my bait tin.

"These highberry larch don't half do well. They fair grow like stink," commented Frank, forcing his words through a mouthful of churning food.

"He! Hee! Aye, yer richt aboot that Frank. They highberry larch certainly do well," chuckled Don gleefully.

Don was at it again. He was taking advantage of Frank. The word "hybrid" did not feature in Frank's limited vocabulary, and he had opted for the nearest word he thought correct: highberry. His attempt caused amusement, which provoked Don to continue.

"Hae ye noticed laddie? Auld Frank disnae carry a bait box like the rest of us – he carries a bloody pantry instead. He! He! Hee! Tell the laddie why, Frank. It'll help wi' his learnin'."

Don exaggerated little in this case. Frank consumed amazing quantities of food and, unlike the rest of us, spent the duration of bait time eating. Insatiable in more ways than one, Frank grasped at the morsel Don dangled, and after a pause for thought turned a face fixed in seriousness in my direction.

"Forestry is hard work lad; yu need to eat plenty o' food. Food contains bitumens, and yu need bitumens to keep yer stanima up."

"Ho! Ho! Hoo! Now laddie – dinnae you forget that," chortled Don. "Bitumens keep yer stanima up. Ha! Ha! Ha! Haa!"

That was not the first time Don had heard the advice, and, judging by his amusement, the entertainment value had not depreciated. It was new to me, and I have to admit to finding it humorous, but I was nagged by Frank's gullibility. Kind and harmless, Frank would help anybody, but Don preyed on his disability and the poor old boy fell for it every time. He deserved fairer treatment. I held Don in high respect, but his ceaseless exploitation of Frank's weakness revealed a weakness of his own.

I was about to replace my empty bait box when I noticed that the spare pullover I carried folded in the bottom of my haversack had gathered an accumulation of needles, sawdust and other assorted dross. As I picked the jumper up with the intention of giving it a thorough shake, my memory was given a sudden jolt by an item that lay among the litter in the bag. It was a map.

A week or two previously old Don had embarked on a journey of reminiscence, detailing his working travels throughout his homeland, north of the border. I had listened with interest to the tales he related, but the places he mentioned meant nothing. At home that evening I had scanned a map of Scotland trying to locate some of the places Don had spoken of, but it proved a difficult task. It took time to translate Don's broad accent into a remote resemblance of the written words, and the pronunciation, in most cases, seemed at odds with the strings of letters. I

100

was intrigued and the incident had given me an idea – the conception of a plot for testing Don, and avenging Frank. That was why I had put the map in my bag. What a coincidence I should be reminded now.

I decided to put the plan into action immediately. With the first place name selected, I split the word into parts, and deliberately labouring over a phonetic pronunciation, I asked: "Don, have you ever worked in Kirk-cud-bright-shire?"

"Ha! Kirkcudbrightshire laddie, Kirkcudbrightshire!" Don said smoothly, with a smile. "Aye, a've worked there."

"How about Stro-nach-la-char; have you been there?"

"Hell and damn it! Stronachlachar laddie!" Don barked, the smile wiped from his face by a dominant glare. "Yer no very guid at readin', ye young bugger!"

The idea was beginning to work; Don was accepting his role, and taking the part seriously. The map on the ground at my side, I moved my index finger on to the next carefully chosen place.

"What's it like at Ball-ach-u-lish, Don. Much forestry there?"

My faltering, puerile attempts at the place names would have embarrassed an illiterate, and my air of innocence was becoming increasingly difficult to maintain. I may have been struggling to suppress my emotions, but Don had no such problem. His countenance flushed with fury; his lips drew into a snarl; his nose wrinkled, and his eyes beamed death beneath a deeply furrowed brow – What fun!

"Ballachulish! Bloody Ballachulish!" screamed Don, re-inforcing the pronunciation with showers of saliva. "Hell, ah didnae ken ye were sic a stupid young bugger! Can ye no get onything richt? Damn ye, ah dinnae believe it!"

Will sat with his head bowed into a supporting hand, and I could tell from the series of shudders that ran through his body that he had entered the spirit of the moment. Frank, alas, hid silent and inert behind a face of stone-cold putty. It was Don, by far, who was the most excited, and he was not a pretty sight. His voice had tailed off into a wail of despair as he balanced precariously on the fence of indecision: fall off on one side and he would explode into a violent rage, and on the other he could slither down the rails and collapse into a whimpering jelly of frustration. I had not anticipated such early results; I had planned for a longer campaign, and still had plenty of ammunition to fire. I pressed on – this was revenge for Frank.

"Drum-nad-ro-chit, Don, have you worked there?"

"Aaaaaaaach! Shit 'n' hell," bellowed Don, toppling off the fence on the side of eruption. "Drumna-bloodi-drochit! Drumna-bloodi-drochit! Ye lang, skinny, donnert, sassenach hoore ye! Ye're no piss wise!"

A picture of purples and reds, Don's face was gripped by contorting spasms of agony. Words were snarled through clenched teeth, steam wisped from his forehead, his whole structure trembled, and knuckles strove to breach white skin . . . Heck! Was he mad!

The success of the plan had exceeded all expectation, and awash with satisfaction, I had only one worry – was Frank appreciating it? Frank's enjoyment would have been my only concern if Don had finished, but he had not. He was struggling to his feet. Drumnadrochit had done the job, but I was disappointed – I still had Auchtermuchty, Gualachulain and Achnaluachrach in my arsenal!

"Ye miserable, lang, donnert hoore ye! A'm gonna kick yer bloody arse till ma boots wear oot!"

The menace was ugly. Don's boots were fairly new, strong and heavy, and, in a test of endurance between them and the target part of my anatomy, I knew which would weaken first! I had never been prone to panic and calmly assessed the situation. Two possibilities existed: I could either make myself scarce, or I could stay. Choose the former, and that would be the end of it, but I might miss out on yet more fun. The other option also offered two possibilities: firstly, Don would deliver his threat, or secondly, a heart attack would get him before he got me – and judging by his colour that was odds on! Having carefully considered all the relevant facts, I arrived at a rational decision – I fled. I could not rely on a coronary to save my . . .!

It had gone just a little too far, and I kept a healthy distance between Don and myself that afternoon. He needed time to cool off. I had discovered that Don could both take and enjoy a joke, as well as hand it out, and his almost instant over-reaction had surprised me. Thinking about it as I worked through the afternoon, I realised where I had made the mistake. I had touched on a sensitive area – an area known as Scotland. In future I would respect Scotland's sacrosanctity!

We had finished the day's work and were almost at the road leading off the estate when I heard Don's voice.

"Come 'ere, ye young bugger."

I looked at Don, trying to glean from his face his intention.

"Here laddie, ah want tae tell ye something – ah'm no gonna hoof yer arse."

Don appeared his usual self; I took the chance and walked toward him.

"Ye lang, skinny, sassenach hoore ye – ye're no donnert! Ye wis deliberately workin' me at bait time, an', ye young bugger, ye made a good job o' it!"

Placing his hand on my shoulder, Don looked at me, his face relaxed with sincerity and said: "There's something ye should know, laddie. There's only twae shits work on this estate – an' you are baith o' them! Ha! Ha! Haa!"

What a relief – Don and I were friends again.

Will and Don moved off homeward on the tractor and trailer, leaving me with a sober-faced Frank, who had just made the gate in his habitual last place. I was donning the top half of a new motorcycle suit, which had come with the Matchless, when Frank stopped by my side. I looked at him, inviting his comment.

"Divvent tak' ower much notice o' thon owld buggor son – some days he just gets full o' hell . . . ower nowt."

Over nothing? – Nothing!

. . . And some fell on stony ground! Shameful waste!

Chapter Nine

The Six Week Sentence

I had not yet adjusted to the speed of the Matchless, and rode into the stable early for what I had been assured would be the first day of six weeks of purgatory. Pulling the bike onto its stand beside the rotovator, I began to have misgivings about my choice of occupation.

One by one a hapless band gathered in the harness-room. Forlorn and silent, its members formed a circle of dejection, awaiting Jeff and his instruction with bated apathy.

"Morning, men. Grand weather!" an ebullient Jeff hailed as he bounced through the door. "We'll transplant the Douglas firs first. Start below the Scots in the second break. Tom, you check over the rotovator, and take Geoff with you; it's sometimes a sod to start the first time."

Conversation was rationed in the nursery, and we filed into the stable without a word being spoken. Tom and I stopped at the rotovator, while the others sorted themselves out with a few grunts and gesticulations. Will and Ron laid their spades on a wooden stretcher and carried it out through the door, heading for the seedbeds, while Don, Frank and The Professor began erecting trestle tables.

The rotovator had been well cleaned and oiled at the end of last season and needed little attention, other than a routine check of the plug, before we were ready to start it. Petrol tap and choke turned on, Tom wrapped the starter belt round the pulley on the flywheel of the large two-stroke engine. Bracing his legs he applied his weight behind the pull and sent the engine spinning over. A further five energy-sapping pulls brought sweat and gasps out of Tom, but not the slightest response from the engine. Tom's exertions had taken their toll, and it was with some trepidation that I relieved him of the belt to take my turn. My apprehension was justified: forces opposed to my efforts united, and at the end of my statutory six-pull stint I was reduced to a lung-heaving wreck that left the engine unimpressed. The resistance of the engine prevailed, and shattered the challengers by the end of round two.

"Hell, I'm knackered!" gasped Tom. "The bugger doesn't want to start."

I was pleased Tom had said it, because I did not have the spare breath. One thing was certain: that engine had more obstinacy than we had stamina.

Petrol was getting through and, satisfied that the problem lay with ignition, we concentrated on the plug. It had been out, dried, cleaned and replaced three times without success when I became aware of The Professor lurking in the vicinity. He did not speak, he simply stood there – watching.

"It's the same every year," Tom growled. "Once it fires a time or two, and gets the slightest bit warm, it'll go and be no more bother after that. It's just getting the bloody thing to fire the first few times."

The plug was out again, but this time for testing. Once it had been pushed into the plug lead of the Matchless, and earthed on the fins of the cylinder head, I switched on the ignition and kicked the engine over. A healthy spark proved the plug was not at fault. We repeated the exercise on the rotovator where Tom watched for the faintest sign of life as I pulled the engine over.

"Aye, it's there, but hellish weak," sighed Tom, picking up the plug to give the electrodes another coating of lead from a pencil. It was then that I began to feel agitation creeping into the shuffles of The Professor as he watched from behind in silence. Turning my head I was not surprised by the expression he wore – I had seen it before. It was that emphatic look of disgust The Professor had cultured over years of suffering the less intelligent – but still he did not speak. He just stood there smouldering with impatience, openly begging a question – I sensed a trap . . . and fell for it!

"Something wrong, Johnny?" I asked, forsaking caution.

I feared the worst, and it followed in the form of a pregnant silence; the mute harbinger of the delivery of a pupil-participating lecture by The Professor. I cursed my curiosity.

"You've struggled on with that thing for the last half hour and you're no further forward. You know what the problem is but you can't see the answer, and yet it's right under your nose!"

Tom and I swapped frowns of incomprehension, and looked back at The Professor, who refused to continue without further student involvement. I obliged: "Eh?"

"Ah wouldn't care, son, but you were so near the answer," wailed The Professor. "Do you never think?"

"Tell me, please tell me," I begged, hoping to avoid protraction.

"It's a matter of common sense, son. Go through the problem a step at a time. First, you know it's not short of petrol – right, son?"

107

"Right."

"And you know it's not the plug?"

"Yes."

"That means it has to be the ignition system – doesn't it, son?"

"It does."

"But the ignition's alright once the engine is warm."

"Yea."

"Well, think back son. You were looking at the answer once."

The Professor never came straight to the point. He revelled in a unique ability to prolong discomfort, and it was preying on my toleration.

"Think about what? For hell's sake tell me!" I pleaded, close to snapping.

"Remember when you put the rotovator plug on the bike?"

"Yes!"

"Think about that, son."

"I have!"

"What did it tell you?"

"That the plug was working!"

"Yes, but what else?" groaned The Professor, grimacing with pangs of exasperation. "The answer's there. What is it?"

I did not know! I was completely baffled, but not alone. Tom was party to my problem, but not equally so. The hint of a twinkle in his eye, and an embryonic smile indicated his advantage of a better understanding of The Professor.

"I don't know, Johnny! I simply DO NOT KNOW, and I give in! Please, please tell me," I cried, the elasticity of my patience tested to within an inch of destruction.

"When the plug was on your bike did it give a good spark?"

"Yes!"

"Well? The answer's there, son. It's obvious if you use your common sense."

I was getting nowhere. I could not see what The Professor was hinting at. His second use of "common sense" had heightened my suspicions of a trap, but, despite being alerted, my brain, tangled in a cobweb of confusion, was incapable of lateral thought. Hauling myself back from the brink of detonation, I decided on a new ploy. Where both gentle and forceful pleading had failed, maybe a total, grovelling submission would work – it was worth a try.

"Johnny, I'm really very sorry, but I cannot see what is obvious to you. I haven't had anything like your experience with engines. If only you'd be good enough to explain, I'd be most grateful, and thanks to you I wouldn't make the same mistake again."

The words had the desired effect. The Professor visibly mellowed, his face assuming an air of benevolence and wisdom, as he stepped forward to impart his knowledge.

"You knew there was a good spark when the plug was on the bike, and that should have given you the answer. What you should do is to put the lead from the bike onto the plug in the rotovator. And if you kick the engine over at the same time as Tom pulls the rotovator over, it will fire straight away. Then all you have to do is to keep kicking the bike over for a few minutes until the rotovator warms up – it's so simple. Yet you couldn't see it!"

The words of revelation were delivered with the rapidity of a machine gun spitting bullets, and they had the same devastating effect. Johnny's suggestion demonstrated an in-depth lack of understanding of engines – it could never work. Petrified, with mouths agape, we could offer no response, and watched stricken, as The Professor haughtily sported a triumphant smirk, then strutted beneath it out through the door.

Ignition timing? – Who needs it!?

TWO – NIL to The Professor!

My grey matter was numbed to the degree where I could not decide whether to laugh or cry, and looking to Tom for guidance I willingly succumbed to his germs of mirth – there would be plenty of time left for weeping, wailing and gnashing of teeth!

With The Professor safely out of sight in the nursery we could ignore his crazy advice and continue with our rustic attempts, Tom once more wrapping the belt around the pulley. A powerful heave produced a grunt from Tom, a cough from the engine and a cheer of relief from me.

It also brought Don and Frank to the scene. "Aye, they're a great invention," Frank commented thoughtfully as Tom rewound the belt. "But they can cause problems – them infernal combustion engines."

Why we laughed I do not know, because Frank was right, absolutely right!

Things unfolded as Tom had predicted. After the first firing the rotovator's fight to remain in retirement was short-lived, and two pulls later we were standing back, breathless but victorious, as the engine whined a high-pitched protest. We had roused a machine that happily slumbered for 46 weeks of the year. But I could understand its aversion to work, when every day of the six weeks work it did was spent in the nursery – that I could sympathise with!

The business of the day, and indeed the next five weeks, was known as lining out, and it was not long before Tom and I joined in with the first part of the job: the lifting of the seedlings from the two-year-old beds. There were two blocks of seedbeds, one-and-two-years-old, each containing fourteen beds, one yard wide and extending twenty-two yards, or a chain. Tom took my mind off the work by explaining that each year new beds were carefully prepared and sown with seed imported from various parts of the world. Selection of a good strain of seed was important, and the Douglas firs we were lifting had grown from seed harvested in the Frazer River region of the USA.

Tree seed germinates slowly, and the maintenance of the seedlings we were lifting had begun seven days after sowing when the earlier sprouting weeds were scorched with the pressurised flames from horticultural blow-lamps. A second similar treatment followed a week later. Once through the surface the tree seedlings were hand-weeded throughout the summers, and protected from snow and frost for the first two years, by which time they had outgrown their allotted space and were in need of transplanting, or lining out.

Unexpectedly, the weather had remained pleasantly ahead of schedule, consistently omitting the cold and wet of the traditional Northumbrian spring. This morning's temperature was in keeping. Heat from the early sun bathed the shirt-clad backs of men stooped to their toil. Soil, only slightly moist, cleanly left the roots of seedlings as they were eased with a spade, gently lifted by the handful and stacked on the wooden stretchers. The morning was glorious and the working conditions ideal, but this was not reflected in the general mood. I could sense the unease, and feel the chill of silence that draped the broken spirits of a gang serving their first day of a six week sentence. Morale was not low – it no longer existed.

Morning bait brought the first respite. Spades were stabbed into the ground at 8.55, and the lifted seedlings were stretchered into the stables. Once lifted, the young trees were at risk of drying out from exposure to sun or wind, and were carried into protection for sorting on the trestle tables.

The nursery was conveniently sited just outside the estate yard. So too, unfortunately, was Jeff's house. Bait times lost the flexibility enjoyed in the woods, the precise duration being enforced by an irksomely effervescent boss breezing through the door of the harness-room. The ten minutes flew. I was sipping at a flask-top of tea, still too hot to drink, when the signal to begin again came.

"Right, lads, let's be having you!"

Those words again, those horrible, menacing words. At the rate of three bait times a day I would hear them another eighty-nine times before release! And this was "interesting"?

Sorting was surprisingly quick. Despite the individual examination of each seedling, the pile of selected plants rapidly mounted. Only the best, fortunately the vast majority, were good enough to be lined out, and any weak or badly formed seedlings were ruthlessly cast to the floor. Tom and I did not get involved; our detail was to prepare the ground in readiness for the lining out, and as we walked back toward the silent and now cold rotovator a question preyed on my mind.

My fears were unfounded. To my relief the rotovator burst into life at the first pull, and engaging forward gear Tom slowly worked across the break. The engine may have been unwilling to start at first, but the machine was good at its job, pulverising the soil into a smooth wake of deep, fine tilth. Four times across and we were ready to begin the lining out of the first plants, which were on their way back into the nursery stretchered between Will and Ron.

"You're going to find this really int'restin', son," Will sighed as he straightened up after lowering the stretcher. "And what's more it'll get more int'restin' by the day. But, a word of warning lad. You may think that the nursery fence is high to keep the deer out, and you may think that the big rifle Jeff has in his office is for shooting them – but you'd be wrong! The fence is there to keep us wretched sods in, and the rifle is for knocking off any poor bugger who tries to escape. It's no coincidence that the office window overlooks the whole of the bloody place!"

Will did like a joke – at least I think it was meant as a joke – but there was no laughter, not even a smile out of politeness or respect for his rank. As a man with supposed leadership qualities, a foreman was meant to be a source of inspiration to the lesser mortals under him . . . This was an example?

I was handed a hessian sack and some baler twine and copied the others, folding and tying the sack on the front of my left knee. Will and Ron buried the roots of the seedlings in a narrow trench to prevent them drying out, while Tom and I strung a line across the break, 11 inches below the last row of a block of Scots pines. Equally spaced across the break, we stood with our spades by the line. I followed Tom's lead, stabbing my spade vertically into the ground against the line, and pulling it back and upward to leave a 'V' shaped trench. With a bunch of seedlings clasped in my left hand I went down on the padded knee and pulled out a plant. I then held the roots at the correct height against the back of the trench, leaving two and a half inches between each root, and covered them with soil pushed in with the free hand. Once the row was complete, the seedlings were firmed by tramping, and the remaining soil replaced and levelled with the spade – and that was it, one row lined out. What excitement! How interesting!

When enough trees had been sorted for the day, Don, Frank and Johnny trudged onto the scene, forming a second gang to our rear.

The morning dragged on. A strained muteness was imposed by the monotony. Only the rotovator vocalised, whining its displeasure as it worked across the break after every second row. The lining out had started about three-quarters of the way up the central break, and at 11 inches a line our progress toward the estate yard was laborious until 12 o'clock intervened.

I was one of a foursome that took a lunch time seat in the harness-room. Don, Will and Johnny who, after making the ridiculous suggestion of coupling the bike and rotovator, had been stripped of his courtesy title had opted for an improvement on sandwiches in the comfort of their homes. Bait time began as always. Flask-tops were unscrewed and filled with tea before coded grimaces disclosed the contents of the opened sandwich tins. Forestry was my chosen work, but in the course of my education I could not help but glean knowledge from the people I met. I had observed that each different sandwich filling induced its own expression which needed no translation. A resigned gasp and a sudden downturn of the mouth, followed by closing eyes and a despondent, sagging head, indicated a dairy product, as in: "Not bloody cheese again!" But if pupils rolled up underneath the eyelids of an open-mouthed, salvation-seeking laid-back head it was sure to be: "Luncheon meat! Bloody luncheon meat! I first had that about six months ago, and I said I liked it − what a stupid mistake! I've had it every bloody day since! Hasn't she heard of cheese?" Whereas it was usually necessary to raise the corner of the top slice of a sandwich to reveal its filling, that was not the case with hard-boiled eggs. They had the endearing quality of introducing themselves by wafting their delicate perfume as soon as the bait tin was opened. Gasps of "Phooorr!" "Urrrrrh!" and "Corrrrr! Who's got hens' farts?" invariably greeted the aroma.

Mouths engaged in eating allowed the usual silence to prevail, but today, for the first time, the crackling of turning newspaper pages interrupted the quiet. I alone had no paper, and was pondering the sudden interest in the affairs of the outside world, when I heard the outer door open. Who could it be at 12.20 p.m.? Imagine my surprise when the missing trio entered. I had not expected to see any one of them more than one minute before starting time, yet here they were – complete with newspapers. Their early return puzzled me. They were certainly not that keen to get back to work! And the newspapers? Nobody had brought one before.

I detected an understanding in the grunts that greeted the three; their entry caused action. Papers were folded and pushed into bait-bags which were placed aside. Seats were shuffled to accommodate seven in a rough circle around an old chest, and Tom, reaching to the top shelf of a battered cupboard, produced the answer to one of my questions – a pack of cards.

It was a game of cards, an everyday lunch time ritual of the six week sentence, that lured men back to the harness-room twenty minutes early. My inclusion in the game was automatic, which was fine until the fourth card dealt to my hand provoked a query.

"What are we playing?" I asked anxiously.

"Knock oot whist, laddie. We aye play knock oot whist," Don replied with a smile.

My fears had grown with every card after the third. With two cards dealt, it could have been pontoon; three, and it had to be three card brag – the only games I had played – but WHIST!? How old did they think I was?

I paid for my learning at the rate of a halfpenny a hand, investing a few coppers before uncovering the strategy and looking forward to the next hand. Clubs were trumps, and I had five: the ace, king, jack, ten and eight; plus the ace of hearts and the king of spades – I could not wait!

Don, on my left, led with the ace of diamonds, which drew a spiralling flutter of lower cards tossed in disgust from the following hands. Perched like a vulture, I awaited my turn, poised to swoop with my lowest club when Jeff crashed in, opening the door behind him – my luck never failed me!

"Right, lads; let's be having . . ."

I had bowed my head and closed my ears in anticipation of those much-feared words. But no. Wait . . . What did he say? And there were still ten minutes left. I looked up, confused, as the hands of cards were placed face down on the table, forsaken in favour of the papers which were retrieved. I only realised what Jeff had actually said when I noticed everyone was scrutinising the same page: the racing page. Jeff had ended his sentence with "the bets"! I should have guessed.

Sixpence each way proved the most popular bet, but Don, true to his ageing but nevertheless swashbuckling style, ventured double that amount with a shilling win on a five horse accumulator. Having little interest in the sport I declined an invitation to join in. Ron abstained also, I suspect because he didn't believe in gambling. The bets written down, Jeff left for the phone and the game of cards was resumed.

My small club trump took the first trick, and I was about to lead with the ace when I hesitated . . . It might be more fun to play the king!

"Och hell! Nae, laddie," Don moaned, as he threw out the three of clubs. "Ye dinnae lead wi' a king when the ace is still tae come."

Surprisingly, I won the trick, but failing to draw the queen, I had to play my ace.

"Ho! Ho! Hoo! Ye young bugger. Yer no so bloody daft," chuckled Don, following with the seven. But where was the queen? I smiled inwardly as it emerged from Johnny's hand.

*   *   *

As I knelt in punishment back at the line, my mind wandered through the events of the day. It was a Monday, the first day of the working week, the first day of the dreaded nursery marathon, the first of gambling on horses, and my first of whist – and no, I did not take all the tricks. Johnny, the ex-professor – damn him – held the ace of spades!

The lunch time entertainment was new, and a complete change of behavioural pattern, but it served a useful purpose as a diversion for the mind. Here we were, seven men with rustic positive blood in our veins, imprisoned in an alien environment. We were instinctive countrymen denied the valued freedom of the outdoor life – any distraction was welcome. Nursery work was not forestry.

Gloom failed to totally eclipse the day; the sun continued to shine – it was after all the first working day of British Summer Time, an occasion I had impatiently awaited. It was a pity that it should be marred by coincidence with something as breathtaking as the opening day of the nursery, but nevertheless I did manage to find the odd moment to celebrate this meaningful date in my calendar. It would bring welcome change in time.

Thoroughly wearied by the repetitious work, I was only too willing to obey Will's instruction to clean my spade and make for the gate after one of the longest days of my life. Soon I would be on the road home with some extra daylight to look forward to, and only 87 more "Let's be havin you's . . ." to go!

*   *   *

117

Minutes ticked into hours; hours chimed into days; days faithfully replicated the drudgery, and expanded into weeks of mind-destroying captivity. The work offered nothing. Granted, we took turns with the sorting, but that gave a change for only a fraction of every other day.

It took the first month an age to idle through the diary, leaving little ink of interest on the pages. I learned nothing more about forestry, but I did gain some understanding of the human body's adaptability – after the second day, the nursery gate triggered the close-down of our brains, and we worked like automatons, giving no thought to the job. Light conversation began to flow, and covered any silly topic except work. Our in-built defence mechanisms had responded, diverting our minds, and making the job bearable.

Don had made an announcement. He intended to work until he reached 70. I welcomed the news, but with reservations. Don was proud and robust, but he had given generously of himself to his many years of work and, sadly, it was beginning to show. I felt he deserved a rest. I did not want to lose my old Scots pal, but the thought of another five self-inflicted years of heavy work worried me.

I spent a few days searching for a tactful way of broaching the subject before I plucked up the courage and said: "Don, when I reach 65 I'm going to retire – definitely! At the minute I wish I could pack in now. I've had enough of the nursery. And you are going to sentence yourself to another five years of it?"

"Och aye, laddie. Ah ken whit ye mean, but it's no as simple as that. If ah retire ah can keep ma hoose, but what would ah do? Ah hav'nae got a car an' cannae afford yen. Ma garden's no big enough tae keep me busy, an' oot here ah'd soon have nothing to do. Ah might as well work. Onyways, if ah work 'till ah'm 70 ah'll get another sixpence a week pension, an' that'll make a difference."

I did not try to press Don. His mind was made up, and I was pleased by the thought of an extended period of guidance under my knowledgeable old pal.

The highlight of the month was undoubtedly the Grand National. Studying the form took a week of serious debate before everyone agreed to differ. The betting was heavier than usual, with each punter backing his favourite, and covering his choice with a bet on two outsiders. Every runner seemed to be carrying money – none of it mine, but I could not avoid the other gamble – a sweepstake. To escape disgrace I had to hand over at least sixpence for one of the folded pieces of paper in the cap. Still smarting from scathing comments about my disinterest in the "sport of kings", I placed a shilling in Jeff's hand and drew two pieces of paper.

I was unlucky! I forget the names of the horses, but I do remember that my first runner fell in the parade ring, and the second won by a nose the following year – but was disqualified when it was realised that it was just finishing the previous year's race.

*   *   *

"Tom, that is one of the most beautiful things I've seen in my life."

"I know exactly what you mean, Geoff – it is a wonderful sight."

"Few people will ever see it."

"Yea, and it will only last for a day or so, but there again, most folk wouldn't appreciate it.

Five weeks into the sentence, Tom and I had just lifted the last of the seedlings, and were taking a brief break to gulp in the splendour of a square chain of bare ground. The lining out was almost finished, with only the new seedbeds to go.

119

"Hell, I'll be pleased when this lot's over, Tom – I'm absolutely sick of it."

"Same here. Geoff, are you serious about going to forestry school?"

"Yes, why do you ask?"

"Because if you don't intend to do that, then you're wasting your bloody time here."

"What are you trying to say, Tom?"

"You're not killed with pay here are you?"

"Not exactly!"

"Well, it's all right you staying here, and suffering low pay to get your two years' practical, if at the end of it you go on to get your forester's certificate. But if, for whatever reason, you don't do that, then you'll have wasted your time here – you'd be far better off contracting."

"You think so?"

"Hell yes, Geoff! You're young, fit and handy, and you're not afraid of work. That's where the money is. Think about it."

My ambition had never been questioned before, and I still believed I was sure of what I wanted to do, but Tom's words lingered to the sweet end of the lining out. I kept seeing Joe, draped in grime and tatters. I saw his sweat fly, his axe flash through its swing, and I saw his shovel full of food. Yes, it was a hard life; but Joe earned more in a week than we did in a fortnight – and he did not have to work in the nursery!

I thought about it. Yes, I thought about it!

*   *   *

The seedbeds were prepared with care. After we had almost powdered a five foot width of friable earth with the rotovator, the bed was marked out with a cord line. Spread along the line, we dug a trench on the outside, casting the soil onto the bed. Beginning at the end of the bed, a dressing of bonemeal was then stirred into the soil by Frank using a five-pronged cultivator. The rest of us, all with rakes, followed Frank up the bed, but in a predetermined order: Johnny next, then myself, Ron, Don, Tom and Will. The object was to remove even small clots of earth, by raking them off into the trench, to leave only the finest tilth in a very carefully levelled bed. The task was progress- ive; the first rake concerning himself only with removal of the larger "knots", but the emphasis changed rake by rake, from "knots" to levelling until Will finished off the impor- tant part. Anyone who could hold a rake was considered good enough to remove the clots, but it required skill to level. Normally, first rake stood on the lowest rung of the hierarchy, which climbed to the elite in the last two places.

My inexperience qualified me as an unknown quantity, and I was positioned where I would have an example to follow – Johnny's! Unknown to myself, I must have had an aptitude for raking, because I was soon promoted from second rake to second from last.

With the preparation over Jeff and Will were ready to begin the sowing. Standing at opposite sides of the bed they each held a cuffer. At first glance a cuffer looked to me like a broomshank, with a plank of wood, about two feet six inches by five inches, nailed to one end. A closer look revealed that the section of the plank was bevelled, like a one foot school ruler, the flat side of which was firmly joined to the shank with angled supporting struts.

Will and Jeff alternated, gently lowering their cuffer onto the bed halfway along its width. A carefully con- trolled pull lifted about a quarter of an inch of the fine soil from the top of the bed and drew it to a mark along the edge. With the bed bared of its covering, Jeff began to sow seed by the handful with practiced sweeps of his arm. Up

121

one side, down the other. The colour of the seed prompted a question:

"Tom, are we growing orange trees?"

"Ha! – that's red lead. The seed is soaked in water for about a week before sowing, and then it's shaken in red lead powder to give it a coating. It's done to stop birds and mice eating the seed – they can be buggers – and at an average of 11 pounds a pound, it's an expensive way to feed pests if you don't coat the seed. And there's another advantage: the bright colour also helps to let Jeff see how thick the seed is being sown."

"Eleven pounds a pound – I'd have to work a whole fortnight with overtime to buy just one pound of it!"

"Yea, Geoff – but don't worry – vermin can eat it faster than that!"

The sowing finished, Will and Jeff began to cover the seed. In turn they reached over the bed, placing their cuffers behind the ridges of soil that had been drawn off. A deft flick sent the soil back across the bed, covering the seed with a remarkably even layer. One bed sown, and I had seen enough.

*   *   *

The weather had held. The sun matched the boredom for reliability and with the work on schedule we had been paroled at weekends, but there was no remission for good behaviour: the sentence had to run its term.

Wednesday of the last week of captivity brought change. Eyes lost their vacant glaze and began to glimpse visions of freedom beyond the fence. Bodies unstooped, faces cracked off their clay casts of misery and words started to flow; normality slowly returned. It was like the week before the school summer holidays when the atmosphere lightens by the minute.

122

A fresh sense of awareness stirred the brain, but the rhythmical motion of the rakes, pushed and pulled at the differing speeds of the individual, had a mesmerising effect – my mind slipped back to my earlier days.

An unusual sight greeted us as we walked out of the nursery at the end of the day. Draped over the gate, giving his best impression of a limp starfish, was Jeff, with his head hung, shaking, between dangling arms. As we neared, he slowly raised his head and faced us without the loss of a single side to side movement.

"I never would have believed it," he wailed. "Never! You buggers – singing, and raking in time! – In the nursery!"

The action of the rakes had stimulated my memories of singing on the school bus in the run up to the summer holidays. It had been a time for rejoicing; a time of song when lungs strained, blasting out glass-shattering decibels, as we hammered through our repertoire.

As we raked I had started singing the first line of a well-known rowdy song. Everyone joined in, and we raked in unison. One rendition led to another, and my backing group performed well, until I came to an old school favourite they did not know. It was a composition of genius; the work of an anonymous northern lyricist, who had selflessly dredged the depths of his talent in search of the poignant words. The title: "Thompson's Father."

I began the first verse as solo vocalist, plastering the words with a thick layer of accent to release the magical emotion:

> Ah knaa Thompson's Fa'ther;
> Thompson's Fa'ther knaas me.
> Ah knaa Thompson's Fa'ther,
> And Thompson's Fa'ther knaas me!

And then the chorus:

> Oooooooooh! – Ah knaa Thompson's Fa'ther;
>   Thompson's Fa'ther knaas me.
>   Ah knaa Thompson's Fa'ther,
> Annnnnnnn! – Thompson's Fa'ther knaas me!

Those who work in the woods are not stupid, and by midway through the second verse I had all of them with me, word for word – until the eleventh verse caught them out. Having not heard the song before, they were not to know that at this crucial point the lyrics changed.

> Thompson's Fa'ther knaas me;
> Ah knaa . . .

That finished Wednesday.

Thursday began with everyone in good voice. Rakes stroked in a harmony broken only by the afternoon activities of the tar-spraying and chip-laying wagons that were resurfacing the South Drive.

Friday opened with the air full of "Thompson's Fa'ther", belted from the hearts of a high-spirited line of men, marching with shouldered rakes toward the last seedbeds – and beyond – to the outside world!

I recall little of that final day: elation fogged the memory, but I do remember going home . . .

The Matchless responded to the first prod on the kick-start. Amid shouted farewells, I let in the clutch to a half twist of chipping-spitting revs before the road caught up with the speed of the rear wheel. Full of exuberance, I left the nursery and its misery behind, and headed home. I knew the road well and spurred the Matchless on, which, to my heady delight, threw distance behind us with contempt. Once over the main crossroads, I accelerated up the gentle curve to the wind-wracked copse that landmarked

Chestnut Hill, where I dropped through the gears, braked and turned right at the junction. Second gear took me along the short straight, and I cranked the bike into a footrest-scrubbing left-hander, before powering out toward the brow of the hill that topped the long, gentle descent to Boundary Burn. A stab with my right foot engaged fourth gear, and with the throttle open, I clawed in the yards to the crest of the hill, with not a single care in the world.

A few words from a small booklet (and one or two others) flashed through my mind as I shot over that hill top . . . toward a solid bovine barrier! The handlebars dipped violently and the back wheel locked as I slammed on the brakes, and fearing I might not be able to stop, I desperately looked for a way through Foster Miller's Friesians on their way to be milked. There was none. My brain raced, searching for a solution, but it was all in vain, and, as it happened, soon out of my hands. I had no choice but to brake, and when braking hard on a bike there is little option but to go straight ahead, which would have been fine if that is what the road had done! It was all over in a few short seconds, but it seemed an age before I ran out of road as it bent to the right. The front wheel left the tarmac and ran onto the dry earth and chippings of the verge and locked instantly – down we went! We hit the grass together but soon parted company, the Matchless wrenched itself from my grasp, as it elected to take an angular bounce back onto the road, and spun toward the cows. Fortunately, the verge was wide, grassy and reasonably level, and I contented myself with a rear-first slide on my back, with feet high in the air. I quickly overtook the bike, which was scraping and grinding its way to a halt on the friction of metal and rubber to road contact. The resistance between the P.V.C. of my jacket and the grass was negligible, and I would have slid to a certain collision with a cow, had not my rear end ploughed into the only gorse bush on that mile-long stretch of roadside.

Thankfully, both bike and I stopped short of the cows. With that worry off my mind, I was free to concentrate on how foolish I felt, as I lay with my backside painfully embedded in a whin bush, peering into the nostrils of an inquisitive cow.

My thoughts were channelled rearwards

"Thoo aa-reet, lad?" inquired Foster.

"Yea, thanks," I replied, rolling out of my predicament, and away from the intimacy of a cow pioneering the kiss-of-life technique.

"Gannin' a bit fast, lad!" Foster politely suggested as he picked up the bike.

"Yes," I agreed, "but I didn't expect the cows."

Foster had the bike on its wheels and was examining it for damage. I appreciated his concern, but my thoughts were channelled rearward. The denim of my jeans was no match for the protective spines of the gorse, and I suffered the consequences. Those parts not numbed by the mass injection, hurt!

"Cows caught ye oot, did they, lad? Aye, it's their forst day oot on grass. Hell, they do like bein' oot."

"I know the feeling Foster: we've just finished in the nursery today."

"Oh, I see, lad; that explains yer hurry," replied Foster, wearing a knowing look. "Aye, a've seen beast gan daft wi' delight, gallopin' 'n' chargin' roond a field when they're forst let oot, but a've nevor evor seen a one travel across grass as fast as thoo, lad – or stop as bloody quick!"

Chapter Ten

Old Woodmen's Remedies

Sitting lightly in the saddle, I followed a newly painted front mudguard to the North Lodge on Monday morning. The Matchless had suffered a rash of abrasions and a bent footrest in Friday's mishap, but fortunately most of its injuries were superficial – more than could be said of mine: those needles had penetrated!

I arrived at 7.25 a.m., and leant the bike against a convenient oak, planted or left by someone with foresight over 100 years before. I walked through the delicate greens of flushing buds, along the grassy ride toward compartment 2b of Park Plantation. Larger woods were often divided into compartments, each given a number. In turn, large compartments could be further subdivided into areas denoted by a letter following the number.

A haphazard, patchy collection of mature oaks and Scots pines dominated the flat, rectangular two acres of compartment 2b of Park Plantation. They towered over a flourishing crop of natural regeneration grown in the absence of man. But that was about to end: our job was to clear the compartment.

This was not another scrub-clearing exercise where

everything ended up on the fire. The oaks and pines contained timber which was destined for the sawmill. I was standing looking at the rhododendrons, the hawthorns, the birch and sycamore saplings, wondering what role I would play in the operation, when the Land Rover appeared. Three doors opened, and six men poured onto the ride, delivering a bright and airy chorus of greetings in my direction.

"Good morning," I returned, amused by a strangeness. Jeff seemed his usual self, but the faces of Don, Frank, Tom, Johnny and Ron were disguised by spontaneous smiles, something I had not seen at 7.30 a.m. for a long time!

I waited at the end of the queue as tools were removed from the Land Rover, expecting at best to pick up my axe and be told to start on the rhododendrons. Everyone except myself held a tool. Don and Frank had their crosscut saw, Tom carried a 7 lb felling axe and so did Jeff, but he looked uneasy with it, and I found out why once he had held an in-depth but wordless discussion with Don. Jeff looked at Don, glanced down at the axe, tilted his head toward me and returned his eyes to Don. Don replied with a positive nod, and the axe was handed to me.

Back out in the woods, and promotion. Great!

"Here, son; you'd better round up with Tom. Watch out for stones and take care of the axe – it's Don's."

Life had suddenly improved, but, after I took the axe, apprehension welled as I noticed a smirk twitching on Jeff's face, and a fire of expectancy in the other encircling eyes.

Jeff continued, "Maybe I should be setting you on cutting whins, son; Foster Miller tells me you're going into gorse bushes in a big way."

I should have known. Word travelled fast. They had heard!

"Aye, but he reckons that you really have a soft spot for

them son – your arse! Hope it's not a sore point. Hah! Hah! Haaaaa!"

The laughter erupted, leaving me to wonder why anyone had gone to the trouble of inventing something as unnecessary as the telephone.

Tom and I began at opposite sides of an oak, working our way round, slicing away the "toes". The oak was large, with matching buttress roots, and we had both started on the same tree to prepare it quickly for the crosscut. Once we had rounded it up, I left Tom to "lay in" the gob and started on another tree. The timber of the oak cut cleanly, but it was much harder than the softwood I was used to, and this justified the seven pounds of steel in the axe head. I had not seen the felling axes in use before; normally four and a half pound "snedding" axes were adequate for the smaller softwoods, but here we were into "big sticks". The felling axes also differed from their smaller relations in shape. They were longer from back to edge, but generally narrower. And they were sharp – extremely sharp.

To have been given this job my prowess with an axe had to have been noticed, and I was both pleased and slightly surprised. I soon forgot the nursery, whin bushes and time. This was fun, not work, and with sweat running I applied effort as I worked into the morning, stopping only to watch the oaks as they fell.

I had grown used to the sight of falling trees, but these were large and mature, bigger than any others we had tackled. The method and principle remained the same, but the effect was exaggerated. The creaks and groans of rending timber as the tree inched away from its place in the sky were louder, but there was no thud of the trunk hitting the ground, just the sharper crash of the shattering branches that cushioned it. I admit to a strange and, yes, exhilarating feeling of power as the trees toppled. For over 100 years those oaks and pines had stood contentedly, interfering with little as they repeated their yearly cycles of growth and rest. Successful and mighty plants, they had shrugged off adversity to reach maturity: a dignified apex in the life of a tree, after which seeding takes precedence and growth concedes. Then we moved in, and demolition took minutes.

Even at my youthful age, the falling trees kindled an awareness of the awesome advantage we appeared to wield over nature. But our powers were not directed against her: we were merely reaping her crop, and influencing her choice of species.

Tom and I together would have struggled to keep ahead of the crosscut if Don and Frank had not helped with the dressing out, which was the relatively slow part of the job. As soon as an oak hit the ground, Ron and Johnny moved in with their lighter axes, removing and burning the smaller branches. The crosscut severed the larger limbs which, depending on size, were either left for the sawmill or cut into lengths and stacked for firewood.

The team was minus Will, who was delivering posts, and my stomach had started to remind me of its existence by the time I heard the sound of the tractor he was driving. We had three trees down, one completed and the other two in varying stages of undress when Will pulled off the ride. It was lunch time at last.

"Have any of you heard about the new saw?" Will asked.

An exchange of glances produced negative shrugs, and a question from Don as our spokesman: "Whit new saw?"

"Jeff's been telling me about it," Will answered. "It's called a chain saw, and it'll revolutionise forestry."

"Whit is it? Whit will it do?" Don queried.

"It's a one-man-operated saw," Will replied. "Basically,

131

it's a small two-stroke engine with handles on, and the engine drives a toothed chain around a bar. It'll make the crosscut saw obsolete – a thing of the past."

"It will replace the crosscut?"

"Yes. All the work we do with the crosscut saw can be done with a chain saw, but a hell of a lot faster and easier. The crosscut saw is finished! The new saw even has a special diaphragm carburettor, which means it can work at any angle – even upside down – just like the crosscut. I tell you, it'll change our work beyond recognition."

Each of us painted our own picture of this wonderful new piece of equipment. Not one of us had picked up even the slightest whisper of it, and it all sounded like something from the pages of a science fiction book.

Don shared our disbelief and asked, "Hell, Will, if they're that bloody guid, why are we no' getting yen?"

"We are gettin' one! It's ordered; we should have it in about a fortnight's time."

The second part of bait time enjoyed an unusual silence as everyone tried to visualise the chain saw, and the implications it would bring. The crosscut saw had reigned without challenge, serving generations of woodmen well, and it was difficult to imagine a replacement that would cause its exile to redundancy.

I was lost in thought when it was time to resume working. My axe swung again, slicing into the timber of the oaks. The sun was hot, and the work arduous, but I gave my all to the afternoon, determined to allay any doubts as to my ability. I need not have worried: no criticism came through the dwindling hours, and at 4.30 I left compartment 2b of Park Plantation thoroughly satisfied – and totally knackered!

If I still had any gorse needles implanted in my backside the following morning, I did not feel them. I suffered agony – but it manifested itself elsewhere!

Six weeks in the nursery, using nothing heavier than a spade or rake, and my hands had softened. To follow that with a day of channelling a pent-up excess of enthusiasm and exuberance through a 7 lb felling axe was asking for trouble, and I got it – excruciation!

The ride to work was painful from start to finish. I could not bear anything to touch the insides of my hands. I could hardly tolerate the gauntlets and I struggled to work using my wrists to declutch and work the throttle.

Slightly late, I walked into the wood with a face that Tom read instantly.

"Morning, Geoff – what's wrong?"

"My hands, Tom, they're killing me. I can't touch anything."

"Not surprised they hurt, Geoff! You put in a hard day yesterday, straight after the nursery. Aye, they'll be sore."

"Sore? They hurt like hell! I can hardly bring myself to look at that axe, never mind pick it up. Tom, I've broken four fingers, the same arm twice and a toe, but never have I experienced pain like this!"

"Ah know what you're suffering, Geoff, but, unlike the breaks you've mentioned, there's an instant cure for this."

"A cure?"

"Yes."

"Instant?"

"Almost."

"What is it? Tell me! I'll do anything – anything!"

"Piss on your hands."

"Get lost!"

Why Tom should want to make light of my predicament, I could not understand, and I made my feelings known by glowering back in contempt.

"Ah'm not kidding, Geoff – piss on them."

"Tom, it's not funny. I like a joke the same as everyone else, but I'm in real pain, and I'll never be able to work today. So be serious for once, please."

"I am being serious, Geoff. Kick the fires in until bait time, and after bait go and piss on your hands. Rub it well in as though you are trying to wash off rosin, and in no time they'll be all right."

Tom was a respected friend, and, although he enjoyed a laugh, he was not noted for an appreciation of sick humour. I kicked the fires in – literally – I dared not use my hands.

I finished bait early; even sandwiches hurt to hold, and turning to Tom, a nod from his expressionless face was enough to persuade me to try the alternative medicine. Desperate, and with nothing to lose but my doubts, I moved off among the trees with suspicious eyes cast backwards – nobody stirred.

Contemplating the idea was bad enough, but words fail me to even begin to describe the difficulty of carrying out the unnatural act. Imagine trying to find a suitable place – when it is always somewhere else, think of undoing your flies while surrounded by invisible pairs of eyes that lurk behind every needle and blade of grass. Picture yourself standing waist-deep in vulnerability, trying to coax an only partially full system to release, when you know that success will trigger an ambush of laughter, catching the defence-less moment of mid-pee exposure, when both hands are thoroughly soaked. That requires concentration – lots of it!

Following Tom's instructions proved a weird experience, but to my amazement there were none of the feared side effects, only an immediate improvement in my condition, and I walked back to the gang feeling both greatly and slightly relieved – and puzzled.

"Done it?" asked Tom.

"Yes."

"Good! Any better?"

"Yea, but I don't understand."

"Give it ten minutes, and you'll be fine."

I worked for the rest of the day as though nothing had happened. I could not believe it! What I had first thought was an infantile trick turned out to be one of the most simple and genuine cures I have found in life. It has been well-tested over the years: I find it always works! A bladder near to bursting point is recommended for the first attempt.

I thanked Tom profusely and offered sincere apologies for my doubts, unaware that further revelations on the subject of rural remedies lay ahead.

*   *   *

After ten days of work we had left a scar on 2b Park Plantation, and almost worn out the chain saw with talk. Soon it would arrive, and give our imaginations an over-due rest.

The "miracle cure" had a lasting effect, and I swung that 7 lb piece of steel throughout the hours without the slightest twinge of discomfort, but on this particular day I was to experience more pain, learn another lesson, and, subsequently, a second woodman's cure.

When Tom and I got ahead with the rounding up, we would stop and help with the dressing out and burning. The fires burned well, but always left a circle of branch ends around their perimeter. Tidiness mattered in those days, and appeared to be affordable. The branch ends had to be thrown in and burnt, to leave the ground clean. Normally, this was a job for bait times when we congregated at the

135

fires, but the sun had refused to take a single day off for weeks, and we tried to hide from heat.

I finished laying the notch into a Scots pine at ten to twelve and as we were ahead of Don and Frank, I decided to spend the few minutes before bait throwing fires in. Feet were useful tools for the job, but not always effective, as they lacked the dexterity of hands. The older fires were just heaps of embers with a fringe of branch ends, but they still gave off considerable heat, making approach difficult, and this is where feet had the advantage. Using your feet, you could turn your back to the heat and keep your head as far away as possible, while kicking the ends in. I had just done that, but my feet could not meet the required standard of tidiness, as they left some unburnt pieces around the edge of the embers. I think it was impatience that drove me to use my hands, and I made a quick grab for the few offending pieces, threw them on the fire, and . . .

"Aaaaaah! ****!" I cursed, shaking my right hand, and for some strange reason limping!

I had a nasty burn and was about to pay a long and painful price for my mistake. My earlier, clumsy attempts with my feet had stirred the branch ends, turning at least one of them round, hot end out. In my hurry I had not noticed and had grabbed a red-hot end. The whole pad of my right thumb was yellow, hard and smelt of burning flesh. And the pain? Exquisite!

My yell had been heard and I sat down to bait with a question from Don: "Burrn yersel', laddie?"

"Yea."

"Bad?"

"Worst I've ever had."

"Och, let's see."

Despite the pain, I could not help but be amused by the aptness of the saying that came to mind as I held my hand out to Don: "Sticking out like a sore thumb".

"Och hell, laddie, it's a bad yen, it'll hurt."

"True, Don – it hurts!"

"Aye, laddie, it'll hurt for days, if ye dinnae fettle it."

"Fettle it? What do you mean?"

"Och, laddie, ye dinnae need tae suffer wi' a burrn. There's an easy cure for a burrn."

"There's no cure for a burn, Don."

"Och aye there is, an' if ye do what ah tell ye, yer burrn'll no bother ye again, laddie."

"You're having me on?"

"Nae, laddie, ah'm no fibbin'. Ah'll make ye twae promises: it'll hurt like hell for a few seconds, and then it'll no trouble ye ony mair."

The imminence of another woodman's remedy filled me with a fear that stampeded my imagination.

"Tell me," I asked, wondering how the nearest cow would react to having my thumb inserted in its anus – or worse, maybe it had to be a bull's.

"Go to the fire, and hold yer burrn as close to it as ye can bear the rest o' yer hand."

"You *are* having me on!" I interrupted.

"Nae, laddie. Forget yer burrn, and think aboot holding the rest of yer hand as close to the fire as ye can stand. The burrn'll hurt like hell, but grit yer teeth, and after a few seconds ye'll feel it start to hurt less and less, until it feels the same as the rest of yer hand. Then it's cured. It only takes ten to twenty seconds. The closer ye hold it, the quicker it works."

A month earlier if I had been offered the cures of bathing my hands in urine or putting them in fire, I would have done neither. However, out of desperation I had tried the urine treatment and it *had* worked.

The sincerity oozing from Don and the thought of

trying to work with a disabled thumb made up my mind. I left the circle and walked to a fire. I had never believed in prolonging anything displeasurable, especially agony, and as I neared the cause, and hopefully, the cure for my pain, I decided to take the shortest route by applying "the closer – the quicker" rule.

Being a good patient, and determined to give the "remedy" a fair trial, I did exactly as the doctor had ordered. Don's cure came in two stages: hell and then relief. I was hardly astonished by the dramatic success of part one, and I concentrated on coordinating my features through a routine of facial gymnastics, while I ignored an exploding thumb! Yes, Don was right: hell was certainly hell! But it was stage two that I sought. I had stuck it out to the point where I was beginning to wonder if I had masochistic tendencies, when miraculously the pain started to diminish. It decreased by degrees, until, as Don had predicted, the thumb felt no different from the rest of my fingers. That did surprise me, and I kept my hand in place for a few seconds more before withdrawing it.

I waited, fully expecting the pain to return as my hand cooled to a normal temperature. It did not. I hesitantly touched the offending area – nothing! Growing in confidence and disbelief, I boldly prodded the injury with my index finger . . . and still nothing!

The treatment had only taken about ten seconds, and I retook my seat as my head was spinning with incomprehension.

"Hae ye din it properly laddie?"

"Yes, exactly as you told me, Don."

"Dis it hurt ye the noo?"

"No, not at all. Thanks. It feels as though it had never been burnt.

"Och guid, laddie. It'll no bother ye again."

Time was to underwrite Don's guarantee. I had learnt another unlikely remedy, rustic, and in this case contradictory to medical consensus, but nevertheless another I have resorted to many times, never to experience failure. It works. Why, I do not know.

It was "lowse" or home time when the Land Rover clattered along the ride and delivered Jeff onto the scene. He brought instructions:

"Right, men. We'll have to leave this job for a day or two. Contractors are coming to thin the block of Sitka in Bracken Glades, and tomorrow we'll have to start cutting rides in for them. It shouldn't be a long job though, the chain saw is coming tonight!"

Chapter Eleven

The Chain Saw

I knew what it was, had some idea of what it could do, but had not the faintest inkling of the role it would play in my life. None of us had seen anything like it, nor even imagined it. We stared, silent, dazzled by its gold and red livery, and awed by the prospect of change – the chain saw had arrived.

The Jo Bu (say it yo bew) had travelled from Norway in a wooden box, together with a tool kit, a plastic funnel and an operator's manual which Jeff handed to Tom.

"Right, men, here it is – the chain saw. It's going to make this job so bloody easy that you buggers will have to start paying the estate for the privilege of working here! Tom, you had better use it. Read the instructions."

Fascinated, I moved in to take a closer look at our new foreign friend with its alarming potential, while Tom followed the instructions and fitted the bar and chain. The initial strangeness began to wear off as I recognised familiar parts beneath their disguise: the two-stroke engine with its chromed exhaust, and the finned flywheel that engaged with the pawls of the hand-pulled starter to send a cooling blast of air over the cylinder; the throttle lever, and the

The chain saw

petrol tank beneath which hid what could only be the versatility-giving diaphragm carburettor. But what were the red lever and the small tank at the front for?

The instructions explained everything. The chain was driven by a sprocket, which was in turn driven by the engine via a centrifugal clutch. At idling speed the clutch ran free, engaging with the throttle only after being put into gear – which was the purpose of the red lever. Oil to lubricate the chain on its passage around the bar filled the small tank, which leaked its contents through a vent opened when the clutch was put into gear – a second function of the lever.

My eyes had been all over the Jo Bu and I understood it well by the time Tom had tensioned the chain and filled the tanks. Eager and impatient with anticipation, I awaited the first sound of the two-stroke engine that would power the new machine into tree-devastating action, and unknown to me, completely change my plans.

We were in Bracken Glades, the largest plantation on the estate. Earlier, on my way in, I had enjoyed a leisurely walk through its varied compartments. The sun blazed, lighting the fresh green of the roadside larches and the delicate fuchsia tint of the small raspberry-like flowers standing along their branches. The pines had exchanged the copper of their buds for the pale grey-green of extending shoots that stood erect like candles on Christmas trees.

The gate opened onto a carpet of grass rolled out down the ride, and the squeak of the catch raised the heads of three roe deer that posed momentarily before disappearing into cover, their white rumps flashing. The weightless form of a red squirrel leapt from the tip of a spruce branch to a larch, silhouetted against the cloudless blue above the ride. Birdsong pierced the heavily scented air, chorusing above the background hum of the wings of countless insects that flew high among the branches in the canopy.

Blissful was that morning in Bracken Glades, but soon the natural sounds of the woodland were to be overwhelmed by the noise of progress.

First, Jeff had to decide where he wanted the rides cut through the solid block of Sitka spruce that filled the slopeless three acres of Compartment 7. Standing on the ride we looked with some dismay at the formidable, solid blackness of a crop that had been planted some 30 years before, and left in apparent abandonment. An unknown quantity faced us. Normally, the 50-foot trees in the 60 by 250 yard block would have received attention years ago, but here we faced a product of neglect.

Trees grow from the top, each season pushing out a new terminal shoot and whorl of laterals above the previous year's growth. The existing laterals also extend their spread yearly until, denied of light, they die, but remain on the tree. Planted at competitive four-foot spacings, the Sitkas of Compartment 7 had raced skyward, chasing the only light available. After 30 years without interference our spruces had grown into a dense mass of trunks with an impenetrable tangle of interlocking, dead and tinder-dry branches.

Jeff had decided to cut two straight rides across the width of the block, dividing it roughly into three. Naturally he sought to choose a route where the poorer trees would be removed, but he had difficulty – he could not see into the block past the second tree!

Normal practice called for brashing to be carried out, either prior to or at the time of first thinning. Brashing, the pruning back to the trunk of the lower dead branches to a height of about six feet, allowed access both physical and visual, and would have prevented the problem that confronted Jeff. I could tell from the agitation creeping into his strides as he vainly paced the ride that he begrudged the price he was paying for the failure of others, and I was not surprised when exasperation loosened his tongue.

"Shit and sod it!" Jeff hissed. "I could walk this ride all bloody day and still be none the wiser. Enough! Right men, grab your axes and come here."

Attentive, and eager to see the chain saw in action, we obeyed the command instantly. Jeff marked a tree.

"This tree is about 80 yards down the ride. Don, you stay beside it. I'll take Will, Tom and Ron, and pace a similar distance down the far side, and I'll shout when I'm there. Right?"

"Right."

No two people plant at the same distance along their rows, and a stagger soon develops across them. Had the rows been planted in the direction of our proposed ride, it

143

would have been a simple matter to remove two or three rows, but true to forestry, they had not been.

"This, son, is where you come in," Jeff smiled with a fiendish glint in his eyes. "When you hear me shout, I want you to start crawling from Don toward me until you get to the middle of the block. I'll keep shouting to guide you. Right?"

"Er . . . right," I replied, looking at the six-inch gap between the lower branches and ground.

"Now, when you think you are halfway in, shout. When we hear you, Don and I will start shouting alternately, and I want you to try to line yourself up between the two of us. When you have, shout, and brash the nearest tree. Then I want you to keep shouting and brash a couple of trees toward me, and then a couple toward Don, while we brash in from the outsides. And, hopefully, we'll get a straight ride. Right?"

That was Jeff's plan, but I could fault it.

"Heck, Jeff! A rabbit couldn't get in there," I protested.

"Maybe not son, but a ferret could – and the only difference between a ferret and you is 90 degrees! It'll be good experience for you!"

Jeff and his crew disappeared, still laughing, round the edge of the Sitka, as I sank to my hands and knees, axe in hand, and tried to peer into the blackness. A shout from Jeff drew a resigned sigh from me as I tried to push myself through the tangle of branches. I soon discovered that hands and knees offered too much body to the rough spruce branches that captured me, grasping my hair, skin and clothing like the tentacles of an octopus. I could not move in that position, and had no choice but to lower myself, prone, onto a bed of well-named needles – or were they hedgehogs? I could not tell because I could not see!

Jeff's shouts encouraged me into an eeriness that held

144

occasion for every phobia: spiders and insects, confined space, disorientation, the unknown and the dark. Fortunately I lacked any instinctive fear, but I did have an apprehension, the result of an earlier experience, and as I struggled blindly into the darkness the memory returned . . .

It had happened one Saturday in November, when I was 14. I had left the bungalow at daybreak with a pocketful of sandwiches, intending to walk the boundaries of my patch. Change never deserted the countryside, there was always something new of interest and I had dallied on my travels. Darkness fell quickly, catching me by surprise some four crow's miles from home. The light vanished completely in a matter of minutes, leaving me groping my way along a fence to the roadside in the darkest black I had ever known. I had no fear of the night, but opted to walk the byroads as the cross-country short cuts would have been slow and hazardous, if not impossible. The glow of the first streetlight on the village outskirts welcomed me onto the main road and illuminated a familiar gate, where I stopped to make a decision. The bungalow lay two fields of ink away in one direction, or almost two miles of lamp-lit detour by road. Hunger and tiredness affected the deliberations, emphasising the folly of choosing the long route when I knew every blade of grass and molehill in the two short fields.

I climbed the gate into home territory, into the steep hilly pasture dotted with large oaks I knew so well. My feet felt for the familiar track that meandered along the edge of the wood to the gate into the next field – but I could see absolutely nothing. Looking upward I could faintly detect the branches of the woodside trees against the sky – the only guide I had to persuade me to force one foot in front of the other along a path I had used countless times, and most recently that morning. Head back with eyes straining skyward, I followed the edge of the wood with arms extended in case of collision with one of the oaks. I was hesitant at first, but had grown in confidence by the halfway mark. I was thinking about food – lots of it – when suddenly my feet stopped . . . I had walked into something on the ground. Something that should not have been there!

Youth carries with it an unshakable belief in immortality and it had served me well up till then. However, for the next few seconds it abandoned me.

"Aaah!" I gasped with shock as, tripped by the unknown obstacle, I pitched headlong into hell. The talons of certain death grasped my sprawling body, injecting an instant paralysis as the world beneath me heaved with a violent eruption. A replay of my young life flashed clearly through my mind, yet strangely everything else seemed to be numbed into slow motion. I felt withdrawn from my weightless form as it was thrust upward and tossed effortlessly into the air. Momentarily it hovered, before being released to fall like a rag doll amidst the commotion below. The sound of nightmare drummed remotely in my ears as invisible action exploded and thundered around me. I thought my end had come. Hell had claimed me.

But as the fanfare of its welcome dissipated into the night, I became aware of a new, and reassuring sound – that of my heart booming out into the pitch as it palpitated in overdrive. I was alive! Slowly I realised what had happened, but having just had the fright of my life I waited for recovery before continuing, very carefully, on the short journey home.

Despite my tiredness I did not sleep easily that night, and next morning I revisited the field of ordeal. Sure enough, there they were – all twenty-three of them. Twenty-three black bullocks that had been comfortably bedded down for the night in a camouflaged friendly circle, until some stupid human had blundered into them – and pounced like a predator onto one in their midst! They had not been there when I set out in the morning, and they left me with a memory I will never forget. Do NOT try this at home!

*   *   *

It was good to be back on the ride, looking into the remarkably straight tunnel we had brashed through the spruces.

"Not bad, not bad at all," said a happy-faced Jeff. "Right, men, let's get on with it."

Don had two outside trees rounded up and gobbed, ready for the chain saw, and we stood, transfixed, as Tom grasped the starter. Four pulls later and Bracken Glades reverberated to the sound of a new era: the angry buzz of a small two-cycle engine that was to become as common as the near-monotone song of the wood pigeon. Forestry would never be the same again.

Thoughts of the future were not in our minds as Tom engaged the clutch and throttled the Jo Bu into the first tree. We had all seen sawdust before, but not like that which the Jo Bu sprayed out as the chain pulled its teeth through the wood. The Sitka fell in a matter of seconds and, despite Tom's umfamiliarity with the chain saw, we were highly impressed with its debut.

We could have remained, locked in thought-induced silence, for the rest of the morning, but Jeff interrupted: "Right, men, what do you think of that then?"

The comments ranged from "Mmmm" to "Aye" – and from Don: "Och it's a hell o' a machine that is. Trust some stupid hoore tae invent it when ah'm aboot tae retire!"

"Aye, you're right, Don. It's going to be the younger

ones like Geoff who'll benefit, but we can't stand here all day thinking about it. Tom, you and Johnny stay here and continue felling. The rest of us will split again and brash in the other ride."

Frank and I followed Don as he paced the 80 yards down the ride and marked a tree.

"Richt, laddie, that's it. As soon as Jeff shouts, off ye go. And, if it's any consolation tae ye, laddie, ah dinnae think that ye look like a ferret – ye're mair like a snake tae me. Ha! Ha! Haa!"

"Bullocks," I replied, simultaneously with Jeff's yell.

"Och, dae ye no mean 'bollocks', laddie?"

"No, Don, I mean 'bullocks'."

"Bullocks?"

"Yes – 'bullocks'." I replied, leaving a perplexed Don in the light on the ride as, once again, I adventured into uncharted jungle.

*   *   *

Filthy, tattered, sore and wilted, we reached the end of a particularly gruelling day. Gravity had pulled down the first ride-side Sitka, but that was the only one that fell un-assisted. All the others were held firmly upright by their grasping tangle of interwoven branches, and had to be wrestled, or pulled free of the very last branch before meeting the ground. Tom's understanding of the Jo Bu had grown with every tree, and its increasing appetite for timber had dictated the pace, forcing our work in an intolerably stuffy heat. There was not a breath of air among the trees. We gasped for oxygen among the dust as we brashed, rounded up, dragged down, dressed out and ex-tracted, trying to keep up with the demands of the new partnership. Sweat flowed, gathering the dust and sticking clothes to bodies. Rough Sitka branches rasped across bare skin and frayed any material that protected it. We stood, sapped of energy, with the salt of perspiration stinging every weal of raw flesh at the close of the first day with the chain saw. The first day with the new machine that was going to make our lives so bloody easy!

"Och aye," sighed Don, "it's been a very clever man who invented that saw, but he's ahead o' his time – by aboot five years!"

By lunch time on the third day we had finished the cutting of the rides, and I had grown jealous of Tom. We had opened the two rides across the block, plus another that skirted the perimeter away from the main ride. The Jo Bu had only been used in place of the crosscut saw, speeding up just the felling element of the work, but I envied Tom his job – I would have worked for nothing but the pleasure and satisfaction of cutting timber with that machine.

"I'm not sorry that's finished," I said, as we sat down to bait. "My arms are red–raw."

"Aye, laddie, they Sitka branches are hellish rough," Don agreed. "An' if ye look at yen wi' some needles on ye'll see why."

Following Don's suggestion, I picked up a branch.

"Well, laddie? Ye ken that both firs and spruce have single needles, but if ye look closely ye'll see that the spruce needles grow on wee stumps, like golf tees, and when they drop off, the stump is left on the branch. Fir needles grow straight out of the branch and when they are cast they leave a smooth branch – a certain way o' telling the difference between the twae species."

I was amused by the accuracy of Don's description. The needles were supported, proud of their parent twig or branch, by a small protrusion that had a remarkable resemblance to a short golf tee, but once the needles were cast they left behind a mass of rough stumps that grabbed and tore at anything within reach. I knew the effect of a branch drawn across bare skin, but later that day one of those dead, but still vicious, branch ends was to find a more sensitive part of my body and demonstrate its true pain-inflicting potential.

Jeff had decided that we should remove the branches and tree tops, or brash, from the two rides we had cut through the block. We had cleared half of the first ride, the tractor and trailer driving in stages over the brash, stopping to allow the following team to grab bunches of branches and pass them up to the "volunteered" stacker standing on the trailer. There I was, staggering about on what looked like a giant woodpigeon's nest, trying to grasp and stack branches as fast as they were offered – an impossible task which I was again assured would be "invaluable experience" for me. Not only was I outnumbered by five to one, but the opposition also had the advantage of walking on level, cleared ground, between trees that had been brashed to above head height. On the trailer I enjoyed no such luxuries. I had the enviable privilege of trying to stack and trample an overwhelming supply of branches that clung to boot-laces, trousers and hair, while staggering amongst the extended branches of the ride-side trees above the brashing height.

Chance controlled my movements as I struggled against the odds atop that heap, but I was given a break . . .

I was caught slightly off balance as the tractor moved forward and, arms flailing, I fought to recover from a backward lurch. "****!" I shouted as my commendable efforts met with acute pain from which I could not escape. My yell brought the tractor to a halt, but I could not turn to see my faithful workmates as they looked up in a silence of concern. It took a few seconds before my predicament became clear, and my onlookers responded magnificently by doubling up and reeling around with laughter. The slightest movement increased the pain and there I knelt, rendered motionless eight feet off the ground, surrounded by mirth-stricken individuals.

At that time, I did not see anything funny at all, but looking back I do concede that it could have been mildly amusing to see someone perched, as rigid as a statue, on top of a trailer full of branches, impaled on the end of a Sitka branch inserted deeply into his left nostril!

The entry of the invading branch was quick, and my natural reaction was to pull away from the source of pain. Unfortunately I was still falling toward the branch, and my reflexes only served to halt further penetration. By the time I managed to regain control of my movements, pain prevented any – I was stuck! The membranes of the inner nose, I had just discovered, were extremely sensitive to intruding objects, and ruptured easily. Dripping blood dampened the humour of my audience, but I was alone because nobody could climb up to help. My attempts at slow withdrawal proved too painful, and were soon abandoned. There was only one answer. A quick, backward jerk of my head, and I was freed of the tethering branch that slipped out as smoothly as a red-hot piece of barbed wire! The solution brought tears to my eyes and left me with the best nosebleed I have ever had!

"Och, laddie, ye're fairly bleedin'," stated a concerned-looking Don as the blood poured.

"Yes," I replied, "I won't do that again!"

"That's guid!" Don nodded, before announcing to all, "The laddie's learnt his lesson. He *nose* no tae dae that again! Ha! Ha! Haa!"

Oddly, but thankfully (I think), there appeared to be no woodman's cure for a nosebleed, and work was temporarily halted until the flow stopped. Jeff took advantage of the lull by wandering off to look for a suitable place for the disposal of the brash. Because of the large quantity of wood, he decided, reluctantly, that a very cautious fire was the only option.

With the trailer piled high, we arrived at the chosen site, a T junction formed by two rides, at exactly three o'clock. Tea was sipped thoughtfully as Jeff delivered a stiff lecture on the need for restraint when feeding the fire. Jeff lit the fire with one match and a bodyful of apprehension. Obeying his command, I walked off to retrieve the miniscule fire extinguisher from the Land Rover. On the way back I was within 50 yards of the fire when I noticed an exception that proves one of forestry's rules: it is decreed that in any block of a single species, there will be at least one rogue. In this case it was a lone Norway spruce which stood, respectfully, slightly apart from its superior Sitka cousins. It grew on an inside corner of the junction, and was the nearest tree to the fire.

As it was late in the afternoon, it would have been pointless for Jeff to send any of us to other work. This left seven of us, plus Jeff, to do a job that could have been comfortably handled by one. Our efforts were normally prodded by a sense of urgency, and I could feel the impatience mounting as we each waited our turn to donate a miserly single branch to the fire. The fire behaved impeccably, and gradually the intervals between our turns decreased, and our growing confidence soon boosted the rate from one branch to two, and then . . .

It was Will who succumbed to overenthusiasm when he cast down an armful that, briefly, appeared to put the fire out.

"Hell, Will!" cried Jeff, as a flicker of dull orange beneath the branches grew into a large bright ball in the centre of the fire. Helplessly we stood as a narrow spire of flame burst through and shot high in the air up the side of the Norway. Only a few horizontal feet separated that hungry flame from the outer edges of the Norway's branches. Had there been the slightest wind . . .

We all held our breath until the column of fire weakened and began to sink to its roots, when relief came – prematurely.

Will's lapse of caution had increased the size of the fire in every way and, as the flames lowered, they widened. It was a relatively small and short-lived tongue of fire that sprang from the edge and licked a taste of the Norway's bottom branches. Mesmerised, we watched as it receded, leaving behind a few offspring flamelets, flickering and dancing through the wispy twigs. Each small flame leapt from birth to maturity, and then died in a couple of seconds, but in every cycle they reproduced. And it took only a further two seconds for the Norway to disappear inside an all-engulfing flame that rocketed from bottom to twice tree-height in a deafening, crackle-filled roar, that scattered us back down the ride.

And then it was time for Jeff – a monument to panic – to remember the fire extinguisher: the nut to crack the sledgehammer!

The extinguisher, of course, was never used. By the time I handed it to Jeff he had realised the futility of trying it, and we stood watching as the flames rapidly died, reveal-ing a stark blackened skeleton. There could not have been more than three yards between the branches of the Norway and its neighbours, but it was enough, on that windless day, to prevent the spread of the fire, something that could have had very dire consequences indeed.

I carried a burden of guilt out of Bracken Glades that night – guilt at being one of only a relative handful of men chosen to work in the woods. Forestry was far too much fun to be shared by so few!

Chapter Twelve

Weeding

Weeding was a job that had to be done, and I soon discovered that little more could be said of it. Newly planted trees needed to be kept free of the more vigorous weeds that would outgrow and smother them. One of forestry's seasonal tasks, weeding was only carried out during summer and early autumn.

We began in High House Gill where, sickles swinging, we each followed a row, clearing a swath through the threatening vegetation. The rate of progress depended entirely on how easily the trees could be found. If a tree could be seen, it took only a matter of seconds to slash through the weed. Time was lost, however, when searching among thick growth for a tree that had been planted out of line, or badly spaced.

During the discomfort of winter's snow, ice, mud and cold, I could remember looking forward to warmer times, but now they were here I was not so sure. The heat I wished for had given us a thorough roasting in Bracken Glades, but now in The Gill we sweltered in the direct glare of the sun. Effort had to be forced out of unwilling muscles, but free was the flow of sweat!

Summer had more to offer the woodman than heat: flies, midges, mosquitoes and clegs – or horseflies – were all part of the bonus. Donations to the blood-hungry were made on a regular basis and, although this could be inconvenient, it was decidedly worse to spend the entire day working with your head buried inside a dense cloud of dizzily swirling flies.

Every fly in Northumberland had gathered in The Gill to greet us on our first morning, and I can accurately describe the experience as a form of Chinese torture. The brain was confused by an interference of vision as a mass of black specks orbited in every direction, and the ears heard only a ceaseless buzzing. It took some bearing. Nothing, I was assured, would act as a deterrent against the flies; the answer was simply to ignore them – but that was easily said.

We were all suffering in our own way, when the nuisance was temporarily forgotten.

"Bastard hot arses!" yelled Frank as he dropped his sickle and took off up the row with such acceleration that he outran his cloud of flies. With mouth cursing and arms flailing, he repeatedly slapped at his neck, head, legs and every other place his hands could reach, while old Don laughed out loud.

"Ha! Ha! Haa! The auld bugger's no' lost his knack," Don exclaimed gleefully.

A casual swipe from Frank's sickle had sent the blade slicing through a paper-like, grey ball suspended in some bracken. This intrusion had instantly stimulated the occupants – wasps, another summer bonus – and, with the noise of his ball of flies jamming his ears, Frank had felt, but not heard, their fury.

"Och, laddie; de ye see that man there?" asked Don. "That man looking like a windmill on piecework?"

"Yes, I do," I answered.

"Well, laddie, I can promise ye that auld Frank has a special relationship wi' wappies – a kind o' love-hate understanding. They love him, and he hates them. And ah'll give the following guarantee: if only one person is stung it'll be Frank. If two or more are stung, Frank will be one of them. And further, Frank will get stung more times than everyone else put together. Wait, laddie, ye'll see!"

I had forgotten Don's exaggeration as we worked into the afternoon. The flies still pestered but I was beginning to acquire some degree of immunity. Wasps, however, commanded a certain respect, and there was an understanding that, should anyone inadvertently violate the insects' privacy, a warning would be given. Searching for a tree on my outside row, I pulled back some shoots growing from an oak stump. I saw the nest at the same time as an injection of venom needled into my right thigh. Fortunately I managed to remember the unwritten rule.

"Wappies!" I yelled as I beat a swift retreat. The nest, I knew, was small and, as I was not being followed, I soon stopped. We all stood still in a staggered line, watching as the aggrieved insects circled angrily, looking for a target to lock on to. None came my way, and I breathed more easily as the leading squadron peeled off in the direction of Don . . . then turned toward Will . . . then Ron . . . Tom . . . and Johnny . . . but finally settled for Frank! Yes Frank! The poor old boy was on the inside row, furthest away from the conflict, yet it was he who was singled out for a successful two-pronged attack – to add to the five he had suffered earlier. At the end of the day the score was Frank 7 – The Rest 1. Amazing! Don had been right.

Lack of rain had subdued the weeds, and we moved on quickly to The Major's, where wasps slipped to second priority. A small compartment, planted with Douglas fir and sitka, it offered a rock-strewn slope to the south. Don and I were on adjacent rows, but it was pure chance that brought us alongside at one particular moment. A stone had rolled onto one of Don's trees, flattening it to the ground, and I just happened to glance across as his left hand was going down to remove it.

"No!" I yelled, throwing out my right foot, which caught Don's wrist and sent his arm in the air.

"Hell, laddie, whit did ye dae that fer?"

"Sorry, Don, but look," I said, pointing to the stone where a large adder lay coiled.

The Major's was riddled with adders: the rocks, patchy bracken and the south facing aspect made it an ideal habitat for them. A private, silent creature, the adder will attack only its prey. In any contact with man, the adder will invariably be on the defensive, but an innocent advancing hand can be seen as a threat.

Whinney Field plantation lacked adders and flies but provided everything else. Wasps, gnats, midges and clegs existed in abundance. True to Don's word, Frank zealously catered for the wasps, leaving the rest of us to deal with the all-female, blood-sucking parasites. Little of the morning had passed before I began to wonder if Frank and I had something in common. Wasps were Frank's speciality but, seemingly, the cleg favoured me. I was introduced to many of the species in a very short time, and after being able to make close observations, I began to ponder the cleg. It is a drab, grey-brown fly, about half an inch in length, which evolution should have tuned to perfection. I would have thought it essential for the survival of any blood-sucking parasite that it can locate its host, land, take its fill and escape without detection. But it is not quite so with the cleg.

Locating its host, it does excellently, and its approach on

silent wings is superb. The landing, too, is absolutely perfect – totally undetectable. But then it makes the often fatal mistake: it bites – and anything oblivious of that cannot contain any blood!

I had decided that weeding and I were not truly compatible by the time we arrived at Thorny Bank Hill. Thorny Bank Farm occupied high, hilly ground on the south east corner of the estate, and it had the largest acreage of pasture land. The fields were unusually big, exposed and bare, save for a haphazard sprinkling of aged, arthritic hawthorns. Beef was the main product, and a herd of about 90 suckler cows had the run of a number of fields through open gates. Two bulls – a Hereford and an Aberdeen Angus – were in attendance.

Thorny Bank Hill was a long T-shaped strip, planted two years previously, to give future shelter on the bleak hillside. But it was unusual in one respect. Situated in the middle of a large tract of farmland, it was remote from a road, and my workmates lived on one side, whereas I travelled from the other. To save the longer ride to join the others I opted to walk the fields alone, and had done so for the past two days. The first three-quarters of the walk was easy – I simply followed a fence, but the last part was tricky. I only embarked, in either direction, after looking for the Angus. It was normal for the sucklers to be split into two herds, each with the bull of their choice, and sometimes they were out of sight. If the Hereford was standing ten feet on the left-hand side of the fence, and the Angus, one thousand yards away on the right, I would walk on the left. The Hereford was like a lovely, white-faced cuddley toy. But that Angus was the blackest, meanest, evilest psychopath that ever breathed fire!

Tired at the end of the third day, I bid my workmates goodnight as we left in opposite directions. Leaning over

the fence, I scanned the fields for sight of the Angus –
nothing. I could see the Hereford in the distance beside the
fence that offered sanctuary, but the Angus . . .? Nowhere
to be seen. To reach safety, I had to cross the 150 yard
width of an infinitely long field, and I had managed one
third of the distance when I heard it: a bowel-loosening
bellow turned my head and filled my eyes with the spectre
of a black death charging on thundering hooves! Adrenalin
propelled me to the only possible haven in that empty field:
the flat roof of a small brick pumphouse . . . and I made it,
just ahead of the Angus. Whether I jumped or flew onto
that eight foot high slab of concrete, I do not know, but I
do remember the relief flooding through my body.

The Aberdeen Angus

The Angus, having been cheated, was mad – blaring,
staring, mad! Although I sat in safety, it was unnerving to
watch the crazed beast at such close quarters. Round and
round he circled, bellowing and snorting out lungfuls of
hatred, his ribcage heaving. Sods of earth flew past his

upheld tail as his forelegs gouged the ground. His eyes flashed out insanity as he tossed his vicious head.

Instinct told me to curl up, showing as little of myself as possible to the enraged adversary, and wait. With nothing to do but think, I had a great opportunity to ponder a question that had entered my mind when I first learned of the Angus's disposition: how could any creature be so miserable when its only chore in life was fornication?

That night I was late home, but wiser – weeding was a pain in the summer!

A Matter of Time

"Hell, it's raining that bloody hard, the fish'll be sheltering under the bridges!" Tom declared, as a dripping queue filed into the sawmill.

Since we had last been there I had asked questions about the mill, prompted by its awkward location, and had learned of its history from Don, and albeit camouflaged by time, the signs of the past were still there to see. I found the tale amusing.

Accident had not caused the mill to be built beside a stream in the bottom of a steep-sided valley. Originally it was sited there to take advantage of water power. The stream itself did not carry enough flow to drive the water wheel, so a holding pool had been dammed upstream.

This turned out to be a huge partial success. Water delivered a remarkably constant power to the saw – much to the delight of all – until lunch time, when the holding pool ran dry. An excellent theory failed, because nobody had calculated that it would take three and a half hours to drain a pond that took twelve hours to fill.

This state of affairs continued for many years until

greater efficiency was called for. It was decided that a steam engine would solve the problem, giving increased output to meet the heavy demand for sawn timber.

Duly the cold water-powered drive was dismantled and removed in favour of the hotter, more modern alternative. A self-perpetuating supply of wood slabs fuelled the fire below the boiler, and Frank was appointed in sole charge of that department.

I am told that Frank took his job seriously, dutifully kindling the firebox every day at 7 a.m. and after much poking and feeding of the flames achieved a working head of steam no later than lunch time – such is progress!

I am assured it never improved – and irony lived until the advent of electricity.

The chain saw underlined its worth outside the mill that day, bringing a new speed and ease to crosscutting. By 4.15 p.m. we had piles of lengths in reserve. Tom and I walked into the mill to put away our gear for the night and found preparations for lowse in hand. A general tidy up was in progress, while Don and Frank, standing at each side of the bench with a slab of wood between them, were crosscutting some nine-inch lengths to take home for kindling.

I bent down for a moment and in that fraction of a second there was a loud bang, a yell, and something flicked my hair on its flight into the upright beside Tom and I in the corner of the mill. I turned, expecting to see Tom on the floor – and he was, but fortunately he had seen the projectile coming and had dived there!

I did not know what had happened, but as I looked back up the mill I saw the pain on Don's face as he doubled up, pulling his hands tightly into his stomach. Instinctively I ran toward him, but horror stopped me dead when I noticed old Frank clamber onto the bench, and stand with blood pouring from his hands right next to the saw, which was still rotating under power – what if he fainted? I dived at the stop button under the bench, and grabbed Frank, pulling him away from the circling teeth.

"That's buggored eet!" winced the old boy, clasping his right hand in his left as Tom and I helped him off the bench. Gently, we eased away the nursing hand to reveal the damage, and one brief look was enough to send Tom to the nearest phone to contact Jeff. Fortunately none of Frank's fingers were missing, but his hand was a mess. A cut spread over the knuckle of his index finger, crossed the next two and turned down the back of his hand to the wrist. I sent him to the first-aid cupboard, intending to patch him up after examining Don's injury.

"Can I see your hand, Don?" I asked.

"Och, ah'm aa' richt laddie. See tae Frank," came the reply which I found annoying. I wanted to tend to each case in order of need, and whereas I knew the extent of Frank's problem, I did not know that of Don's.

"LET ME SEE YOUR HAND!" I ordered, with a new-found authority and a fixed glare. Don immediately complied and exposed the back of his left hand. There was no break in the skin, but it looked as if there was a black hard-boiled egg, implanted between the knuckles of the first and middle fingers. Satisfied, I went to Frank.

The large flap of skin that once was the back of Frank's hand folded back to open a beautifully clean wound. I could see veins, bones, tendons and blood, but not the slightest sign of dirt. Frank was clearly shaken, but there was no sign of any broken bones, and I could do little but bandage his hand in an attempt to close the wound and staunch the flow of blood until he reached hospital. The flap of skin was loose and after a few vain efforts I realised it was a four-handed job, two to hold the dressing in place and two to apply the bandage.

162

I was vaguely aware of Don in the background, pacing in pain-induced circles when frustration took over.

"Hell! I'm sorry, Frank," I apologised, "but I haven't got enough hands."

Don was there, offering assistance, before the words were out of my mouth, and with one of his fingers, one of Frank's and my two hands we managed a respectable piece of first aid. With nothing more to be done, Frank, Don and I took a seat and waited for Jeff, and it was only then that I became aware of something strange: at the time of the accident, there had been seven men in the mill. Tom had immediately gone to inform Jeff – which left six – but where were the other three? Gradually it dawned that since Tom had left, I alone had tended the injured, and when in need of help, it was Don who had assisted. I could not recall seeing Will, Ron or Johnny – they had mysteriously vanished!

"Don, where are the others?" I asked, suddenly aware of their absence.

"Probably ootside, laddie. Will cannae stand the sight o' blood, an' the other twae are no much better – as ye can see!"

When I heard the Land Rover coming I escorted Frank out of the mill, where I saw the three missing persons. They stood separately, with heads bowed, shuffling their feet through their own discomfort.

A brief lack of respect had caused the accident. In their haste to cut a few lengths for kindling at the close of the day, Don and Frank had relaxed one of the rules. Instead of placing the flat side of the slab onto the bench they held the rounded face down with their fingers underneath. Easier to grip in this way, the slab was being rapidly reduced to short lengths when the last cut was fed, hurriedly, into the saw – too hurriedly. The teeth were not given time to cut, and

163

grabbed at the timber, rolling the slab onto Don's left and Frank's right hand, between which it was smashed, and not cut into two. Don suffered for about a fortnight, but Frank was given six weeks of time for reflection – confined to a hospital ward.

Forestry was a tough, unforgiving occupation – accidents awaited – it was only a matter of time.

Paying the Price

The sight of Hades surprised me! I had always imagined it in red, but the one laid out in front of us beckoned in every shade of green.

Five weeks had passed since summer had been snuffed out by a sky so overburdened it wrung out its clouds in panic. Ceaseless rain had washed us into the sawmill for a fortnight.

Gradually the heavens had tired, and reduced their output to spasmodic showers which allowed us back in the woods, and reminded Will of the nursery – and weeds. Although it was wet, it was warm, and the fact that the periodic rain kept us out of the nursery perturbed me little. But the same could not be said of Will, and I admit to alarm at the deterioration in his mental state as we squelched through the days in our wellies.

Will's bouts of delirium began to infect us all, and by the end of the fourth week, growing anticipation of the inevitable plunged us into depression. It was five weeks after the downpour began when the words struck like a knife through our hearts:

"Right, men; it's dry enough for the nursery tomorrow – there's weeding to be done."

Plucked from the despair of expectation, we were

dropped into the hell of realisation as we looked aghast at a sea of weeds held in by the nursery fence. With not a tree to be seen anywhere, it was an exact reproduction of the picture Will had painted when lecturing me in my innocence.

"Aye," groaned Will. "Ye sure will find this int'restin', lad. Very int'restin' indeed!"

I did not reply. I just wanted to run – from those words I could see on my headstone!

There was no thought of singing as we folded and tied sacks around our legs that morning. The prospect of three weeks grovelling on our hands and knees pulling out weeds killed that, and closed down our brains for the period. There was a price to pay for neglect.

Hopelessness kicked a new enthusiasm into the bait time diversions. Cards were played with haste, in order to cram more games into the time, and the wagers placed on horses increased to almost double! Don, who studied the form of the horses more than anyone, had been impressed by the consistent improvement of a horse by the name of Xerxes, and amusement came on the day Don decided to back it:

"Och aye, and ah'll hae a tanner each way on that Sex-rex back'ards."

Lighter moments were few but welcome, and helped to spur us through the gloom and into the second week when my mind took off on a flight of mental escape. I thought of contracting, agriculture, and even banking. Anything would be better than this – yes, even stonepicking!

As I worked I remembered my first, and last, experience of stonepicking. I had asked Farmer Watson for a holiday job and he had driven me, in a Jaguar to a huge grassy field.

"Here we are, son. Grab those sacks and I'll show you how to do the job."

With a bundle of hessian sacks under my arm, and a mouth open wider than the gate, I followed Mr Watson into a field that grew as I scanned it. Rolling down the top of a sack, he handed it to me, saying, "Don't carry it, son. Drag it behind you and throw in the stones as you go. Start off here and keep about six feet off the wall. Pick stones six feet to your right as well as those to the wall and you'll cover twelve feet each time across. Understand?"

"Yes."

"The next thing you have to remember is not to fill the sacks too full. As soon as it gets slightly heavy, stop and empty it in a pile on your right, marking the edge of the ground you have covered. Keep doing that 'till you reach the wall at the end of the field. Have you got that?"

"Er — yes," I mumbled, searching the distance for the stone wall.

"On the way back, aim for six feet on the high side of the piles of stones you left. Try to keep the heaps in lines up and down the field, but remember, don't try to drag a heavy sack — empty it. Understand?"

"Yes, I understand," I replied, heavy with resignation, as Mr Watson turned and made for the gate.

Suddenly the enormity of the task owerwhelmed me. There I stood, a solitary 12-year-old, walled into a 20-acre field covered with a rash of stones. A youngster dwarfed by a challenge that would have sent a gang of hardened prisoners fleeing to the protection of their cells! However, I had asked for work, and been given it by a thoughtful farmer who showed concern that I should not strain myself dragging heavy sacks.

"Do a good job son, and call at the farmhouse for your pay when you've finished. But remember, don't fill the sacks too full – they wear out too quickly if you do!"

Suddenly I felt as limp as the sack I held. If this farmer

was worried about the cost of a few sacks, what was he going to pay? When looking for work there was no prearranged agreement, it was a matter of trust, but this farmer was an unknown quantity and I was apprehensive.

Fourteen weary days of picking abrasive stones had worn every trace of a fingerprint from the hand I held out for my pay. But amazement compensated, lighting my face as Mr Watson unpocketed a roll of money, and peeled off seven pound notes into my hand: ten shillings a day. I had never seen so much money!

"Yes, son, you've done a good job. Any time you want some work, call in. I'll always find something for you. And thanks."

From the moment I had first seen Mr Watson standing in the doorway of his farmhouse, I realised he was not a typical farmer. His wavy, silver hair was carefully groomed, his suit was tailored from fine material, and his glasses had delicate metal frames instead of the usual thick black rims. His polished Jaguar and soft, well-spoken voice removed him further from the average – as did his unequalled generosity. Mr Watson was a true gentleman farmer and I spent many days at West Field Farm, but strangely my visits seemed to coincide with busy periods when no time was spare for stonepicking!

Visions of that field and its stones lingered in my mind, despite the rich reward, and haunted me for years. Nothing could be as bad as stonepicking, I had thought . . .

*   *   *

The beginning of the third week in the nursery brought into view the end of the green misery, and our spirits began to lift. Stunted conversation on next week's freedom cheered us further and I was starting to become more optimistic when Tom spoke: "Have you heard, Geoff?"

"Heard what?"

"Ah'm leaving."

"Leaving? You can't!" I protested.

"Well, sort of leaving. Ah'm going Next Door."

The news was shattering. Two of my six colleagues were more than workmates – they were friends, and Tom was one of them. Initially I did not relish the idea of work without Tom at all, but the full story changed my mind.

Tom had been promoted. A lot of contractors were working Next Door in an attempt to wrestle as many acres as possible from nature's grasp over a five-year planned period. The supervision on both estates was more than Jeff could manage and Tom had accepted the newly created foreman's job. I was pleased for Tom, he knew forestry and deserved to do well.

But it left a question: who would operate the chain saw now?

Chapter Thirteen

A New Appointment

It had to be an insect! That I knew, because the body was divided into three distinctly separate parts: a head, thorax and abdomen, and the thorax had the qualifying three pairs of legs attached, plus a large pair of wings. Yes, it was an insect, but of such proportions it could only have mutated in the imagination of some horror story writer! An alarming two inches in length, with a black and yellow head, black thorax, yellow legs and antennae, and an abdomen banded in the same bi-coloured warning, the insect mimicked a wasp, but with subtle variations. This weird creature sported a short pointed appendage at the business end of its abdomen, but it further underlined its menace by advertising a black, inch-long sting slung below its body. The sting emanated from the front of its abdomen, and continued backward to protrude as the ultimate threat.

The detail is accurate: I had an advantaged view. It had landed on my elbow and was making its way up my shirt sleeve in jerky steps to my shoulder . . . I did not move . . . I dared not!

Clegs were accepted as a nuisance and wasps respected

for their sting, but they did not intimidate me – however, this thing . . .

"What the hell is this, Don?" I yelled in a whisper of fear.

"Whit laddie?"

"This, on my shoulder!" I said, with eyes crossing in the direction of the all-to-close alien.

"Och, laddie, that's a wee wood wappie; it'll dae ye no harm. It's no really a wasp, it's a wood-boring fly that is looking for some wood, usually larch, to lay its eggs in. Ye've nothing tae fear – unless yen lands on yer heid! Ha! Ha! Haa!"

"But the sting?"

"It's no a sting, laddie, that's an ovipositor – it's whit it bores into the timber wi'. Keep still and watch it."

Calmed by Don's assurance, I relaxed and began to take an interest in the new super-insect.

We were baiting on a heap of freshly fallen larch logs, and the smell of the rosin had attracted the visitor. Spattered with sap, I must have confused the insect initially, but it quickly realised its mistake and flew off onto one of the logs. The flight was short and I watched closely as the creature drew itself along the rough bark with hesitant, mechanical leg movements. Only inches away, I could see the body gradually arching, lifting the thorax higher until the "sting" parted to reveal the protruding point of its brown, needle-like ovipositor that slid along the bark. I watched, intrigued, as the insect continued arching and then went into reverse. As the wood wasp went backwards, the point of the boring apparatus caught in the bark, pulling it down and out of the protective sheath I had mistaken for a sting. The rearward motion stopped when the ovipositor was almost vertical, and the insect's legs at full stretch. It looked like a miniature drilling rig.

I knelt and watched, fascinated, as the boring commenced. Hooked feet clung to the bark, pulling down on the ovipositor that appeared to consist of four separate, finely toothed sections that moved up and down in turn. The power came from abdominal muscles that tirelessly gyrated with an artistry that would have shamed a belly dancer. Gradually the ovipositor shortened, and the body lowered. At a guess, it must have taken between one and two minutes for the insect to bore to the correct depth, deposit an egg through its ovipositor and withdraw, before moving on a few inches to repeat the reproductive chore.

I felt privileged to have witnessed one of nature's more secretive occasions in such detail. The wood wasp, horntail or *Urocerus gigas*, as the entomologist would call it, is seen by few as it lives a secluded life in coniferous plantations. Although the insect is completely harmless to anything other than trees, this fact is belied by its colouration, spectacular size and generally fearsome appearance. And I have heard of accounts where one of these creatures, dissatisfied with the peace of the woods, has flown off and through the open window of a crowded bus in search of excitement – and I can imagine the fun it had!

My first encounter with the wood wasp is still clear in my memory, but it will not linger as long, or as vividly, as my experience of the first day in East Pit Wood.

With a weekend to readjust to normality after the nursery, I arrived eager to get back to forestry – thinning – amongst real trees once again. Life had restarted!

We had assembled, as arranged, at a crossrides that divided an extensive block of larches into four unequal compartments. All were present except Tom, and I distinctly recall feeling the hollow left by his absence: he was going to be missed.

With the gear unloaded from the Land Rover, a

thoughtful silence fell as Jeff looked at the joiner-made box that contained the Jo Bu.

"Well, son," he said, "it looks like you are the chain saw man now. There it is; get on with it."

Words elude me to express my delight at the realisation of a coveted ambition, and the fact that I had gained it by default did not detract from my elation. Both Tom and I had won.

Red tape, in the form of training, did not exist in those days, and when the chain saw had been handed over to me I embarked on a course of self-education. I made mistakes, learned from them, and enjoyed every minute.

Northumberland, it is said, enjoys nine months of winter, and three of bad weather. The exceptional spring had lulled us into a false sense of security, but that had been rudely shattered. Summer failed to recover from the rain-lashed fortnight that had kept us in the mill. We had days of rain, and the odd week of sunshine, but during the brighter periods intermittent showers could be relied upon to provide a weed-sustaining moisture. Our movements were as unpredictable as the weather. The sawmill, the nursery and the outside, divided our attentions. It was a time of unpredictability, and change.

My new appointment brought unexpected benefits. I was elevated to a status where, if chain saw work was needed and a full team was not required in the nursery, I was spared the indignity – a great improvement, and only one of many.

On one of the days in the sawmill we had been surprised by the arrival of Bob, the estate mason, and his assistant, Jim. We rarely saw them and assumed that they had called in to collect some timber. Puzzled faces watched as sand, cement, bricks and metal bars, pipes and sheeting were barrowed into the mill.

"Whit in the hell are ye doin'?" asked Don.

"Don't you know?" replied Bob.

"Nae, ah dinnae."

"We've come to put your fire in."

"Och! Away tae hell ye lying hoore ye."

"I'm not having you on, Don. You're getting a fire."

"Hell, ah dinnae believe it! A proper fire? – The joiner makin' oor armchairs? He! He! Hee!"

The news came as a surprise to us all; not even Will knew of the plan to modernise the sawmill. Two days later the back wall of the mill boasted a fire with a huge heart and a matching mouth to feed it. There was no doubt that the bait times of the future were going to be warmer, but the main attraction of the installation was a metal pipe that rose from the top, climbed the wall and turned outside below the roof . . . it had a flue!

As predicted, the coming of the Jo Bu had left the crosscut saws hanging in retirement on their nails on the sawmill wall, but progress was not to stop there. New equipment seemed to be delivered each week. A brand new diesel Fordson Major tractor, complete with a modern winch, was first to arrive and make the old faithful redundant. Two more new machines quickly followed: a saw-bench and a peeler that were to confine the outfit to the sawmill yard, sheeted in obsolescence.

Both the McConnel sawbench and the Cundey peeler had their own wheels and diesel engines which gave independence from a tractor except for when they were being moved from site to site. The twin-cylindered engine of the McConnel gave it plenty of power to drive the 30-inch saw, but the main feature of the bench I had not seen before. It had the usual rectangular table of a push bench for ripping, but this could be folded down allowing a long narrow table, at right angles to the saw, to swing into and back from the blade, for crosscutting tree lengths.

The peeler was much less demanding of drive than the sawbench and had a more than adequate single-cylinder engine. In principle it worked like the old machine, but the Cundey had been designed by someone who knew the job of peeling. A powered cogwheel, with variable height and speed, supported and helped turn the timber against the blades – a great advance that reduced effort, and the risk of peeler's wrist.

Forestry was leaping ahead in efficiency, but I was soon to realise that the new machinery did not make our lives any easier. The work was just as hard – but our output increased.

New equipment always generated interest, if not excitement, and our sudden influx prompted Will to recall the memory of a hay time in the past.

Arnie Walton was a cautious local farmer who thought carefully before parting with money. Balers were new to the area and Arnie had, adventurously, splashed out and purchased one. Unsure of its performance, he had tried not to advertise his acquisition. The baler was red, tractor-drawn, and had its own engine, which was to turn out to be prone to gathering hayseeds and other bits and pieces. Everything worked perfectly and Arnie had finished baling most of his hay after a week, by which time he was bursting to tell the world.

"What's it like then?"

"What's what like?" Arnie replied nonchalantly.

"You know bloody well – your new baler!"

"Oh that . . . it works like a dream," Arnie proudly announced to a Friday-night pub full of sceptical agriculturalists. "This is the way forward," he insisted. "But don't take my word for it. Come to the ten-acre field tomorrow afternoon, and you'll see it working."

Arnie had circled the field a number of times by the time the first spectator turned up, and he continued lapping until the Cushat's Nest pub emptied. Then people started to arrive. Arnie was pulling a well-warmed baler by the time he gathered a full audience and, when satisfied with the demonstration, he stopped the tractor and walked toward the arc of onlookers.

"What do you think of that, then?" he asked, confidently.

"I'm impressed," said Miller Herdman wearing a puzzled smile, "but does it always catch fire?"

Arnie whipped round, where to his horror he saw flames leaping from the still-running engine of his new and precious toy. Panic animated Arnie's every action.

"Aaaaaaaah! Quick, get the fire brigade – get the fire brigade!" he screamed, running in a number of directions at the same time. The spectators, unlike Arnie, were un-affected by the excitement of the occasion; nonplussed they stood there – watching.

"Do something! Do something – don't just bloody stand there," pleaded a hand-wringing Arnie, hopping from foot to foot. "For hell's sake do something!"

"Don't panic, Arnie; calm yersel' down," reassured a cool Miller as he stepped forward and walked toward the flaming baler undoing his flies as he went. "Ah'll put the buggor oot fer ye!"

To the great delight of the spectators, Miller proceeded to hose the flames with a well-directed pressurised jet of processed beer.

"Never worry, Arnie; ah'll soon put the buggor oot!"

And he did! To Arnie's great relief – and to rapturous applause!

With winter approaching I, too, thought of a change of machinery. The Matchless was a good bike, but it was too small to solve the problems of the coming snow and ice.

The Speech

I had traded in the Matchless for a B.S.A. Gold Flash. With its twin–cylinder 650 cc engine it had the power I needed. My idea was simple. A year's road tax could be bought in three four-monthly periods, and this would allow me to tax the Flash in the sidecar/cycle class from the 1st of December through to the end of March. I intended to buy Tom's spare sidecar chassis and leave it in the garage until snow or ice called for the third wheel. It would only take minutes to fit, and with a bag of sand as ballast I would have transport I could not fall off. That was the plan, but for later in the year.

Most of the summer had been spent thinning, but now in early autumn we were clearing a large area of birch, which was to be planted in the coming season. We were hard-pressed for time, highlighting a snag I had discovered earlier.

The Jo Bu and I worked well together; we had developed an understanding and I loved the work, but being the chain saw man had a drawback. Overtime was occasionally compulsory and more often than not, available if we wanted it. If work was behind I accepted the extra hours, but it was the voluntary overtime that irked. Saturday afternoons and Sundays were not too bad; it was the extra hour and a half added onto the end of the first four days of the week that bothered me.

At that time I was almost 18, and although I looked on forestry as fun there was more to life than work – the female!

By the time I had travelled home at the end of the extended day, eaten and had a bath, half the night had gone. This nuisance curtailed my evening activities, but I felt obliged to work because of my position. Older and married, my workmates had different priorities. They needed money and always volunteered for the extra six hours the four longer days gave them. Being the chain saw operator, either I worked, or nobody did.

Since being awarded the Jo Bu I had suffered in silence, not even giving Don as much as a hint of my feelings, but slowly my annoyance grew. It was not so much the overtime that I objected to: it was the pay that really hurt. At seventeen I was three years behind the top rate of pay and overtime that everyone else received, and yet I was doing a job that they relied on. It did not seem fair that I should have to work unwanted hours, doing the most important job, for a fraction of the rate of others who depended on me. No, it did not seem fair, but I said nothing, and continued.

Life was full of surprises, and I was to sample two of its extremes on one of the days when we were clearing the scrub off Chestnut Hill. After a three-week relationship I thought I knew the Flash, but I had neglected one of its requirements which it alerted me to on one of September's brighter mornings. The Flash was contentedly purring at a sedate 70 mph up the long gradual incline toward the crest of the hill where I had met Foster Miller's cows when travelling in the reverse direction.

I shut off the twist grip as we neared the top of the hill – but to no effect. And, once over the brow and onto level road, the Flash accelerated – the throttle slide had stuck!

It took a few fractions of a second before I realised what had happened, but that was long enough to fill me with alarm. My stomach baled-out at the sight of the quickly looming, 20 mph right-handed corner, and my heart fell into its place! I hit the cut-out button with my left hand, my right grabbed the front brake lever and my left foot stamped on the pedal that locked the rear wheel. Together the bike and I slithered along in partial control, but our deceleration could not match the speed of the oncoming fence! The posts and wire demonstrated a remarkable resilience, bouncing back without harm after stopping the bike in mid-skid . . . but I continued, over the handlebars, and landed flat on my back in a field full of Foster Miller's cows! I was not injured, but I had two major worries: had Foster seen the incident? And, where was that paramedic cow? I quickly left the scene!

Yes, it gave me a fright – and I added Redex to the petrol in future. I was getting to know the Flash.

Slightly subdued by the adventure, once safely at work I started to mow down the birch for my workmates to slice into manageable lengths and burn, and as the Friday continued, the memory of the morning paled.

"Right, men, gather round. I want you all to hear what I have to say," Jeff announced when he arrived with our pay.

We left our work and stood around Jeff wondering what on earth he was going to say – and nobody was going to be more surprised than I was!

"Something has been bothering me for a while," said Jeff, "and at last I've been able to do something about it. As you all know, young Geoff is doing an important job – he's the chain saw man, but he's paid a lot less than you – and that doesn't seem fair. Therefore, it has been agreed that he should be given a six shilling a week rise. And, furthermore, he works overtime when a lad of his age has better things to do, but he doesn't grumble. He gets on with the job which allows you to work the overtime you want. And for that reason we've decided that he should receive the same overtime rate as everyone else. Any objections?"

Not a word was uttered. I was speechless – but suddenly I felt appreciated – and that was heartening. What a day!

*   *   *

Joe was a regular visitor. He had been working Next Door since before Tom had left, and often called in to update us with the not-so-local news. His appearances were always appreciated and usually he carried a tale of interest. One I particularly recall centred on Eric.

Eric had moved in with Joe. He was one of Joe's younger cousins, but, in spite of his 20 years, Eric had a special quality of innocence. An apparent underexposure to life had left Eric short of knowledge – and, on occasions, Joseph short of patience!

Saturday night was Joe's highlight of the week, and he went through a strict routine of preparation before embarking on an expedition in search of ale and the opposite sex. Joe's transport was a motorcycle combination, which questioned the time spent grooming his hair. However, Joe thought the effort necessary and was combing the last locks into place when he had a thought.

"Eric, can you check the bike for petrol please. We'd better be safe than sorry."

The back door closed behind Eric as he walked out into the yard. Two minutes later a loud "boomph" rattled the kitchen window and tore Joe away from his reflection. He shot outside to find a stunned Eric sitting astride the bike, his left hand clasped the petrol cap on the tank, and in his right he held a spent matchstick – but at first, Joe did not see the match – or Eric's lack of hair.

"What the hell happened?" Joe asked.

"It shouldn't have done that . . . it shouldn't have . . ."

Eric's plaintive wail was cut short as Joe spotted what he was holding in his right hand.

"You bloody stupid bugger! You lit a match and . . ."

"But it shouldn't have done that," Eric feebly interrupted.

"Idiot, bloody idiot! Ah wouldn't have believed that anyone . . ."

"But Joe. It shouldn't have done that."

"Hell! Ye're a bigger fool than ah gave ye credit for. Ye've got to be the only bugger in Northumberland that'll strike a bloody match to see into a petrol tank – and then wonder why it went up! And what do ye mean: 'It shouldn't have done that'?"

"But Joe, that's what ah've been trying to tell you," pleaded Eric. "It was a *safety* match!"

Prison Again

The idea of the Flash and its stabilising wheel was simply to make the winter journeys easier. Falling off and dragging your feet on snow could be both embarrassing and tiresome. The combination matched my expectations, but I had not foreseen the added bonus.

Winter had provided plenty of snow for trial runs, and I soon got used to the new technique of handling a combination. But more than that, I also found its great potential.

Other vehicles were a rare sight on the country lanes I travelled, especially when snow made the main roads more attractive. This gave me a freedom to unleash the bike and chassis on some of the clear stretches of road . . . and what fun I had! Not being able to fall off gave me a new confidence to challenge the slippery surface and use it to advantage. I travelled many an exciting journey that winter, powering the bike through corners in (usually) controlled slides, testing nerve and machine to the limit. Although the rides left me frozen stiff, I enjoyed the winter so much I was sorry when the snow began to thaw.

But I mourned the passing of the winter for another reason: it made way for spring – and the nursery. I have to be honest, and admit to being the chief nursery hater. My workmates were older and likely to remain in their work until retirement . . . I was not! They accepted the nursery as a disagreeable part of the jobs their houses tied them to. But I was not tied. So why was I about to sentence myself to prison again for another six-week stretch, beginning tomorrow? Even I could not understand that.

I had learned that Tom and his solitary workmate, John, were being conscripted for the period. Whereas I would be pleased to work with Tom again, I did not wish the circumstances of the reunion upon him.

"Hello, Geoff; surprised that you're still here," said Tom.

"Nice to see you again, Tom," I replied.

"Never expected you'd go through this again, Geoff. Thought you'd have seen the light and gone contracting. This is John. He'll tell you about felling timber."

I shook hands with John, a man in his fifties with a sinewy, medium frame, but with an immediately noticeable jovial personality that shone. The sharp features of his rubicund face were affable; a pair of eyes radiated a mischevious glint through the thick lenses of his glasses.

"Thinkin' oboot gannin' fellin', lad?"

"I have thought about it."

"Well ah'll give ye a bit o' advice. Watch them timber morchants – they're buggors – an' ye'll hev te l'arn te stick up fer yersel . . . Ah can mind the time when ah was workin' fer Brownie and he come aroond one day and stood on the stump o' a tree ah'd just felled. 'Look at this stool,' he complained. 'What's wrang wi' it?' I asked. 'It's far too high. Don't you realise that the timber merchant gets his gold from the very bottom of the tree?' 'Aye – ah knaa that,' ah telt him. 'So fetch yer bloody spade an' dig fer the buggor – 'cos we poor sods ownly get the coppers off the top! Aye – ah telt him!' "

I liked John.

Ron was a totally different character. He took life seriously. I did not often work with him, but when I did I found him difficult. Unlike the rest of us he seemed to spend more time trying to keep his hands and clothes clean than he did working – a strange priority for a woodman. He did not use the normal channel of releasing frustration, either – he only swore at Christmas trees. But, worst of all, he was not a creature of habit. And that was to lead to a problem.

Some days, especially when we were working in the nursery, Ron would ask me to pick him up at the North Lodge the following morning, and give him a lift up the North Drive. This mattered little to me as it made no difference to the length of my journey. Routine is easy: it is a mechanical rut we readily fall into; it is variability that proves difficult. Ron was one of a rare breed that thrives on unpredictability. He would ask to be picked up one day in one week, two different ones the next and none the following. Had I picked him up every day, it would have been simple.

I think it must have been during our third week in the

nursery that I made the mistake. By then my brain had slipped into defensive hibernation, and it was only when we were trudging into the nursery to begin work that I realised my slip

"Hell, I've forgotten to pick Ron up!" I blurted out to Jeff. "Can I go and get him?"

"No, you can't," came the stern reply.

"But I said to call for him."

"It's up to him to get himself here – and on time. Get on with your work, or you'll be in trouble too!"

I was looking at a Jeff I had not seen before. His usual good cheer had been ousted by a foulness he could not hide. Subdued, I got on with my work, next to Don.

"Och, laddie, dinnae tangle wi' yon bugger this mornin'. He's in a hoore o' a fettle."

"So it seems, Don. Never seen him like this before – what's happened?"

"Ah dinnae ken, laddie. Me'be his horses ran back'ards yesterday! He! Hee!"

I could never explain the depth of the guilt I felt as a forgotten Ron traipsed through the nursery gate. It was like watching a newly hatched chicken leaving its shell and walking straight into the entrance to a fox earth where the predator waited with saliva drooling from its newly sharpened teeth. I cringed.

"And what the hell do you think you're doing trailing in here at twenty minutes to eight?" Jeff bellowed at Ron.

I could not take it. It was my fault, and I could not stand aside and allow someone to take the blame for a mistake that was rightfully mine. I left my work, and intervened.

"Don't blame Ron," I pleaded.

"Shut up!"

"But it's my fault!"

"Keep out of this!"

183

"I was supposed to pick Ron up, but I forgot. That's why he's late – you can't blame him. Play hell with me – it's my mistake."

Jeff released Ron from the fury of his glare and slowly turned to me.

"Did you hear what I said, lad? I told you, shut up and mind your own business. I meant what I said – and if you don't stop interfering and get back to your work – you'll get the bloody sack. This has nothing to do with you. Keep out!"

"But it's not fair!"

"I'll be the judge of that, son – get out of my sight. Now!"

I could do no more. Jeff, for some unknown reason, was beyond talking to that morning, and I went back to the job with the sound of Ron's totally unwarranted bollicking ringing in my ears.

An uneasy silence prevailed in the harness room at bait time. I felt awful, and each icy glare that Ron cast in my direction froze me further in my guilt.

"Sorry, Ron, I just forgot," I said in an attempt to break the deadlock. "And, when I remembered, Jeff wouldn't let me go to get you. I apologise – what more can I do?"

"You're not a man!" snarled Ron.

"What do you mean?" I asked cautiously.

"If only you were a man . . . but you're not!"

"I don't understand – what more could I have done?"

"If you were a man you'd have told him the truth!"

"I did; I told him it was my fault!"

"But you didn't tell him what really mattered!"

"I don't know what you mean."

"If you were a man you'd have told him that it wasn't twenty to eight when I got here – it was only twenty-five to!"

184

My guilt flew off in amazement leaving anger in its place. I had not been looking at my watch when Ron crawled into the nursery – and that was enough. From that moment on that bald-headed, cigarette-sparing, non-swearing, dirt-conscious, unappreciative idiot, was on his own. Sod him!

My youthful flare of temper had lain quietly, long-forgotten in dormancy until that particular incident re-kindled it – now it smouldered, sustained by boredom – awaiting a breeze.

* * *

Jeff's mood had oscillated between bad and impossible over the five weeks we had been interned, and I could read the evil in his strides as he burst through the gap in the hedge near the seedbeds, and headed straight for me. There was no mistake. We were all more or less together, but at a distance of 50 yards, a half glance was enough to tell me that Jeff had singled me out for attack. I waited, wondering . . . and he confronted me.

"Right, now you've gone too far, son! You are far too clever with a pencil for your own good. You should have kept it in your pocket. But now, the writing on the wall is for you!"

I knew instantly what had upset Jeff. Some individual, in search of momentary escape, had obliterated the word "Tsuga" from a name plate in the seedbeds and had replaced it with a more relaxed "Sooger", the phonetic version, but I was not guilty.

And neither was I deserving of the automatic reprimand that flamed the embers of my temper. I lost control, I saw red – but the language verged on blue!

"Go to hell," I snapped.

"Do you not realise that foolery like this is . . . What did you say?" asked Jeff, taken aback by afterthought.

"You heard what I said," I told my accuser in a clear voice. "I told you to go to hell!"

That made Jeff erupt, but having lost my temper I was ready for anything. The sack – or whatever – it did not matter! I was not going to allow anyone to trample over me because of a false assumption. That, I would not take.

Standing eye to eye with Jeff, I was prepared for every eventuality when a voice interrupted: "Maybe it wasn't him, Jeff," shouted Tom, his conscience prodded by the point of a pencil I knew he possessed.

The words dropped a curtain of silence both around and between us. I stood my ground facing Jeff – waiting. Slowly, his eyes fell away and his head followed – and without a further word he strode out of the nursery.

I believe he recognised his error, and understood my reaction, but it sickened me further – and made me think of contracting.

Chapter Fourteen

Further Education

I stopped instantly beside a tree, slid my body behind it and froze. There in front of me, curled up asleep in the bracken, lay a large dog fox.

It was a warm, sunny lunch time, and after emptying my bait box I had decided to have a look at the *Abies grandis* that we were going to thin next. With only the wisps of a gentle breeze in my face, I had been walking through the patchy bracken when I stumbled, almost literally, on the fox. I held my breath and watched.

Although he was asleep, the animal's senses must have been alert to some degree, and have fed their information to a slumbering brain.

The animal's eyelids began to twitch, and then parted to reveal a pair of glowing amber slits. Lethargic blinking further opened the eyes as a raising head slowly revealed the fox's muzzle, drawn from below his brush. He paused, in half slumber, awaiting the coming of a gaping yawn that uncurled his body and heaved it into a sitting position. Still blinking, his eyes scanned the surroundings, but I suspect little registered – he did not see me. A discernible reluctance straightened his back legs, raising his rear above the

187

ground where he stood momentarily before being gripped in a spasm of stretch. His spine arched, pulling together legs that quivered as the stretch climaxed. Suddenly he relaxed and his front legs walked forward leaving his back pair stretched out, flat along the ground behind him. Then he stood, looked around once again and walked off to disappear into the bracken . . . totally unaware.

I was indeed fortunate to catch one of nature's more wily animals in a rare moment of vulnerability, but that was an example of the broad-based education forestry was to provide as the months flipped through the calendar.

Time developed preferences in me for some of the tasks forestry demanded. I liked working with trees only when they were yielding timber, and I enjoyed converting it into finished products. That gave a satisfaction that justified the effort. But planting, weeding, fencing and ditching, although essential, failed to inspire me.

We always seemed to be behind schedule with thinning, and the constant demand for mining and fencing timber kept a gang busy, either thinning or converting or both, on every available day. The Jo Bu was heavy, noisy and vibrated, but it performed superbly and my love for it grew as it shepherded me away from some of the less favourable jobs. Both the McConnel and Cundey quickly established their value, speeding up the sawing and peeling, but they did not ease the arduousness of the work because we simply left larger heaps of props and posts behind at the end of the day – and this was to have repercussions.

Whatever the job, there was always an expected target for each man per day. Planting asked for 600 trees, splitting larch posts 200, and so on, but there was always leeway to allow for varying conditions, and work was hard, but free from pressure. It may have happened with the course of time, but I suspect that it was the introduction of the four efficiency-improving machines within a relatively short period that brought about change. Our production output increased dramatically, almost overnight, smashing the old targets into sawdust. Jeff, of course, noticed and erased the old figures from his memory – but it was when we realised he had replaced them with new, limitless ones that pressure entered our lives. Forestry was in the early throes of change.

In all my thoughts on forestry I had never envisaged the word "commercialism" creeping into a sentence on the subject. The honest nature of the work, and its rural setting, far removed it from the infection of the demands of business and industry that existed in the cities . . . or so I had believed in my youthful innocence.

In fact, financial exploitation was already well cultivated in one area of the estate – the nursery, yes the nursery – and that explained a lot. I had learned that the nursery was the best-paying entity on the place, and that was why the fence entrapped three whole acres – far more ground than was required to supply our own needs. We were growing trees to sell at a profit each year, and with interesting results.

Estate agents ran the estate, and took many of the orders for produce, but, blinkered by the thought of money today, they did not appear to consider the future. Consequently, after a good year's sales from the nursery, the estate was often left with a remainder of trees that did not match its own planting requirements. Norway spruce were planted on sandy hillsides, and Scots pine in wet land – all totally against forestry practice, but with no option because we could only plant what we had left.

I could not see the sense of short term gain taking precedence in a job that could exact at least 40 years of retribution for a mistake in the choice of species planted, but that is how it was – disappointing.

The Jo Bu introduced me to the *Abies grandis*. Not often planted, the grandis is unusual for another reason: it is generally referred to by the specific part of its Latin title, and its common name, giant or grand fir, is rarely heard. Its English name is more descriptive, giving an indication of the huge size it can attain when grown in favourable conditions. Distinctly geometric, the grandis pushes out level whorls of branches that carry horizontal rows of shiny, dark-green needles. The bark is generally grey – and blistered – the main feature of the grandis. Only two of the almost 900 acres of woodland on the estate had been allocated to the grandis, highlighting the unpopularity of this fast-growing tree. At first it seemed strange that a tree which grew quickly to large proportions was neglected to such an extent. However, the dubious quality of the timber discouraged the forester – and the blisters sickened woodmen.

All but the youngest and oldest bark was liberally studded with blisters; easily broken blisters that squirted a highly aromatic, sticky liquid rosin at the slightest provocation. It was unavoidable. It dripped, oozed and shot in every direction, and at the end of the first day we stood, glued into our clothes, reeking like highly concentrated toilet fresheners! Two acres was far too much!

An oval, chromed metal box fitted the exhaust port of the Jo Bu, and I admire the imagination of the person who, according to the handbook, called it a silencer! It did little to dampen the noise that pounded the eardrums and breached the peace of the woodland.

It would be fair to assume that the presence of men working in a forest would alone be enough to disperse wildlife from the area. And I would have expected the added decibels of the two-stroke engine to banish every creature from within a square mile, but that was not the case, and normality soon returned.

190

Lunch time stopped the Jo Bu, allowing silence to return. Ron had walked home, leaving me perched alone on a ride-side heap of twelve-foot lengths of timber with my back against a larch. Only ten minutes of bait time had passed, and I was quite surprised when I saw, about 50 yards upwind, a roebuck leave the cover and walk into the sunshine on the ride. He stood proudly, with head high, showing off his unusually red coat against the lush green of the grass. I stopped eating and waited, motionless. Being downwind and silent I had the advantage; he could only detect me with one of his senses – sight – and I knew how blind the roe was.

He neared, a few paces at a time, stopping to nose at the grass for a second or two before impatiently throwing aloft his head and jerking his eyes around him. But he did not look directly at me, and kept coming . . . twenty yards . . . ten . . . Less than four yards were between us when instinct unsettled him, and after an agitated wide-eyed search he noticed something of interest – me!

He did not flee; he had no reason to because he could not interpret what he saw. My clothing was drab, and my normal upright human outline was camouflaged by my position and lack of movement. He stared, first with one eye and then with two. I could feel his brain trying to shuffle the image into a picture as his head tilted to one side and then the other, bobbed up, then down but all to no avail.

The roe had nothing to fear but man, and to have watched this normally wary creature walk to within a few feet of his only predator and then stare at it was beginning to amuse me. I knew, once I felt the first twitch of a snigger, that more would follow and betray me . . . and after spending almost 20 minutes disguised as a statue I could not allow that.

"Oh dear deer!" I boomed in my best schoolmaster's voice. "Not very alert today – are we?"

"Berrrrrrrrhh!" he bellowed in reply. At last he had made up his mind, and sank low to the ground before springing away with a powerful bound. I could contain myself no longer, and released my pent-up laughter as he shot off down the ride for a few galloped strides, veered off and disappeared into a patch of tall bracken. The undergrowth sheltered a track that was well-used by the local deer, and – coincidentally – by Ron who was ambling back to work.

Imagine the surprise that lurked in wait along that twisting, secluded track as the two parties closed – Ron at a leisurely two mph, and the roe in full flight! "Flight" took a more literal meaning when the two met head-on! I heard a simultaneous "Aaaaagh!" from Ron, and a "Baaarrrff!" from the startled animal as it leapt high into the air. Well clear of the bracken, I was given a brief, but spectacular, view as the roebuck sailed gracefully over Ron's head, to land out of sight with a short "Baarrh!" that was repeated with every few strides, and decreased in volume as the animal left the track behind it with urgency.

A shaken Ron arrived and relived his experience. "Ah don't know who got the biggest fright!" he said.

Looking at Ron I was not sure either, but having rolled off my perch on the heap of timber, I knew who had enjoyed the biggest laugh!

Characters

"Was there an accident at your place last night, Johnny?" Will asked, opening the conversation as we sat in the mill sheltering from the rain. "Ah saw some broken glass on the road this mornin'."

192

"Yes, there was," replied Johnny, the ex-Professor.

"What happened?"

"Let me see . . . it must have been after half past six, because I'd just seen the news on the television – did you see about that murder?"

"Yes, Johnny, we did – about the accident?"

"Oh yes. It must have happened just after the news, because I'd gone for a shave . . . an' aren't them Wilkinson Sword Edge blades a great improvement? They last much longer than – "

"Yes, Johnny, we know that – the accident?" Will pressed.

"I remember now. I was shaving when I heard the bang. I was just going to look out through the window when that crack in the pane caught my eye. Does anyone know when it's going to be replaced?"

"No, we don't, Johnny. Tell us about the accident!"

"Like I said, it must have been soon after half past six, because the news had just finished. I was thinking about that murder while I put a new Sword Edge blade into my razor, and I'd just started to shave when I heard the bang. I stopped shaving, and when I looked out through the cracked window toward the crossroads I saw it . . ."

"Well? What did you see?" begged Will.

"Firewood – at last my load of firewood had been delivered. How long is it since I ordered that?"

Will gave up.

Johnny could entertain at length, volunteering information on any scientific topic of his choosing, but when asked a direct question or to recount an event, digression possessed him. It was rumoured that Johnny knew it could be highly stressful to be greatly overqualified for a job, and had settled for a life in the woods rather than follow his aptitude to Westminster!

Johnny had lived in the South Lodge for 17 years. The cottage was near to a junction on a busy byroad, which undulated and twisted, between the main road and the nearest village. With two more junctions within a mile of the Lodge, there were the occasional accidents. Johnny, when it suited him, would fill us in with the details, relevant or 99 per cent not, which always stressed the importance of his involvement . . . and I do remember one particular story . . .

Sitting in the sawmill one bait time, we were collectively suffering a usual painful account of a recent incident. We had almost dozed off on the detour when Johnny finally neared the point, and suddenly gave me the opportunity for revenge – one back at 'The Professor' – and I grabbed it!

"It was a young couple in a sports car," said Johnny. "They were travelling towards the main road, an' had just gone ower the brow at Whinny Rigg, when suddenly the road was full o' sheep coming towards them. They were doin' aboot 75mph, an' – "

"Heck! That's fast for sheep!" I interjected, grasping my unexpected chance.

"No! No! No! Not the bloody sheep," snapped an indignant Johnny, as everyone rocked with laughter. "The people in the car, ye silly young buggor! Hell ye're stupid!"

"Oh, sorry," I apologised. "Thought they might have been Lambhorghinis – or even Maaaseratis!"

The joke was a welcome release for all but Johnny, whose countenance flushed through a spectrum of disgust. After all, it must have been difficult for him to put up with someone as ignorant as I was . . . poor Johnny.

*   *   *

Bill was tall, fresh and the best company that any man could have wished for – so long as work was not expected of him. He had "worked" on the neighbouring estate as horseman until the new horsepower of the tractor had awarded his partner retirement – something Bill had unofficially enjoyed for years.

His boyish face underestimated its 50-odd years by at least ten. His head was topped with black hair that curled tightly around the sides of a green cord cap. His warm, appealing brown eyes invited conversation, and he returned it – with interest. Bill was blessed with the remarkable ability to speak incessantly for eight hours, and then continue into overtime without pausing for breath. His conversation was interesting but unrelenting, and he frayed many a tough pair of ears!

Apart from his face, Bill shunned description by secreting the rest of his person in wellies and a large army greatcoat that swaddled his body in khaki, permanently, regardless of season or weather. However, I am assured that there was one memorable day, sometime in the nine years he had been on the estate, when the impossible had happened. Allegedly it had taken a temperature that hovered at around 90 degrees in the shade, the diversion of a general anaesthetic and the local blacksmith's oxyacetylene burner – but Bill had shed his coat!

Well ahead of his time, it is now thought that he was the true founder of the energy conservation campaign – the only movement he condoned – but Bill was not a selfish man and applied his principles fairly. One example springs to mind.

Clad as ever in his coat and wellies, Bill had spent most of a winter morning standing in a foot of snow, happily splattering any available ear with a ceaseless barrage of words. Meanwhile, patiently tethered to a tree, his horse had succumbed to the cold and began to stamp its feet in an attempt to restore the circulation . . . but to Bill, that was squandering energy.

"Stan' still! Stan' still ye stupid buggor! Ye'll tire yer bloody sel' oot."

Bill's physical actions may have been slow but his brain was quick, and he had chanced upon a useful ruse soon after moving onto the estate.

On one of his early visits to the nearest village, he had trudged the two miles of country lane that brought him to the main road. He travelled in search of provisions and some paraffin for which he carried an empty gallon can. Turning left down the main road, Bill had walked little of the further three miles when a car pulled up beside him.

"Run out of petrol?"

"Err . . . Aye!" replied Bill, suddenly realising the significance of the can in his hand.

"Hop in; I'll give you a lift."

From that day on, Bill always carried a can to the village, whether he needed paraffin or not, and he soon perfected the deception. "It pays to advertise" was Bill's first lesson and he quickly found that the most effective method was to wave the can at arm's length, while jumping up and down in the path of vehicles going his way. Most times that worked! The outward journey's only requirement was an empty can, but if no paraffin was bought he could not pretend to be returning to a stranded car with an unfilled container. Being resourseful, Bill solved the problem by acquiring a second can. One he kept for the paraffin, which he took when needed, but the other was carried on the dummy runs. It went to the village empty, and came back containing water — secretly filled in the horse trough at the rear of the blacksmith's shop.

Bill's plan served him for many years, saving him miles of walking. It even worked for a second and third time with the same drivers, who probably suspected, but enjoyed his company. Like many strategies, however, there was one flawed element, and Bill's most embarrassing moments came when overgenerous motorists insisted on taking him all the way to his car!

Word had it that the estate planned to replace Bill, when he officially retired, with a tortoise. It was thought that one would be faster, quieter and more easy to coax out of its khaki shell!

Yet everyone liked Bill. He was a character.

*   *   *

Work gave me the opportunity to meet but few of Northumberland's scattered wealth of characters. Of those I had the pleasure of knowing, one couple in particular will linger forever in my memory.

I was honoured to discover Johnty and Betty. Together they tended a 20-acre smallholding by day, but on Saturday and Sunday nights one of the local hostelries was certain to be graced with their presence – and the inevitable entertainment that came with it.

Johnty fizzled with charisma; Betty did not. Johnty worked in a pair of clogs and a tired cap; Betty chose wellies and a colourful apron. At night their footwear rested beneath the apron and cap which relaxed on a nail five feet up on the back door. By day the cap perched the same five feet high atop a knotted wire called Johnty, while the apron strings strained to reach round Betty. Thick waves of lily-white hair covered Betty's head, highlighting the pleasant but quiet face of a tall and buxom woman.

Johnty had a few long hairs, a pair of large, black-rimmed spectacles and little else. The glasses fought an endless battle for grip on what could be seen of a slender nose. Thick lenses magnified his eyes, their blueness and honesty, to the point where the rest of his features dissolved. Short and slight was Johnty, so much so that gravity had spent over half a century ignoring him – but not so Betty. Betty had what Johnty lacked. They were a complementary couple, renowned for their flair as entertainers.

A double act, they could hold a crowded bar enthralled for hours listening to their tales. Nobody uttered a word when Johnty was speaking: drinks were ordered by nods. The pair would enter a bar and sit together in silence like a ventriloquist with her dummy. Betty with her gin and tonic in front of her, Johnty disappearing behind his pint – waiting for a cue. Johnty needed little excuse and it was never long before one of the locals would oblige:

"Your cold any better, Johnty?"

"Naw – ah canna get rid o' it! That right, Betty?"

"That's right, Johnty."

"But ah'm not surprised. When ah lie abed aside Betty the blankets don't touch me. It's like sleeping in a wind tunnel. No wonder ah've alwis got cold. And ah'm alwis getting wet. That right, Betty?"

"That's right, Johnty."

"Aye, ah remember the other mornin' . . . ah got soaked then . . ."

Here Johnty took one of his occasional, well-timed pauses, inviting audience interaction – "What happened, Johnty?"

"Huh. It wis aboot fower o'clock in the mornin' when ah hord a hell o' a commotion amang the hens. Ah dived oot o' bed an' grabbed me gun. That right, Betty?"

"That's right, Johnty."

"It wis just bre'kin' daylight and fair stottin' doon wi' rain when ah ran ootside in me nightshort . . . Then ah saw the buggor – the biggest dog fox ah've evor seen in me life. It wis the size o' a bloody wolf! That right, Betty?"

"That's right, Johnty."

"He saw me at the same time, an' ah must ha' given him a fright 'cos he dropped the six hens he had in hes mooth, and he wis off, gannin' like hell towards the byre. That right, Betty?"

"That's right, Johnty."

"Ah lifted the gun as fast as ah could, but the crafty buggor thowt that he'd getten away. He was oot o' sight, weel aroond the corner when ah fired, but ah still got the buggor! Aye, he wis roond the corner but ah got him! That right, Betty?"

"That's right, Johnty."

"Aye, eet's one hell o' a gun, that new un . . ."

"Got a new gun, Johnty?"

"Ah have that, an' eet's the best un ah've evor had. When ah forst browt eet back, ah saw a big buck rabbit sittin' away up the field. It wis a lang way off, but ah thowt eet wis worth a try, an' took a shot. Does thoo knaa, ah killed eet stone deid – at 100 yards. An' when ah went to pick eet, ah wis amazed. Eet wasn't ownly deid – eet wis gutted, skinned an' half cooked! That right, Betty?"

"That's right, Johnty."

"Aye, eet's a hell o' a gun that un – but ah've sowld me fower-ten, 'cos ah'm not bothered wi' rats any more . . ."

"No rats, Johnty?"

"Not any more. Ah've got the best rat-catcher in the fower northern coonties. That right, Betty?"

"That's right, Johnty"

"Rat-catcher, Johnty?"

"Oooh aye – wor Sally – the shorthorn coo . . ."

"What, the milking cow you recently bought?"

"That's hor. The wisest animal ah've evor knaan. That right, Betty?"

"That's right, Johnty."

"But how does a cow catch rats?"

"She doesn't catch them – she kills the buggors! That right, Betty?"

"That's right, Johnty."

"But how, Johnty?"

". . . Well, ah'll tell ye . . . Ah've watched hor . . . Like ah said, she's the cleverest beast ah've evor seen. She hates rats an' soon worked oot what the buggors were up to. They came in through the wall from the hayshed, an' the ownly way into the byre wis through a hole above hor heid, but there wis ney way doon for them . . . Sally knew that an' . . . she'd wait until she knew a rat wanted to be oot . . . An' then she'd slowly raise hor heid an' press hor nose onto the wall just below the hole . . . The rat, seein' a way oot, would walk oot alang hor heid while she neithor blinked nor breathed . . . Doon hor neck it'd gan – an' onto hor back . . . She'd nevor move until the rat wis halfway alang when she'd evor so slowly raise hor tail an' hold eet straight oot, but slightly to one side . . . The rat, wi' neywhere else to gan, had to walk alang hor tail . . . But she'd wait . . . an' wait . . . until it got to the end . . . And then, wi' a flick o' hor tail, so quick that the eye couldn't follow, she'd clash the rat . . . stone deid ag'in' the wall! That right, Betty?"

"That's right, Johnty."

"Does thoo knaa; when she first l'arned that trick I spent many a mornin' barrowin' deid rats oot o' that byre! That right – "

What imagination! Brilliant!

Chapter Fifteen

Promotion?

After two years on the estate, if I had been given the choice of changing any one thing it would have been the age difference between Don and myself – I would have made Don younger. I had the greatest respect and admiration for my old pal, but a gulf of years separated us, and I often wished I had known Don in his youth.

In his younger days Don had played as hard as he had worked, and spent every Saturday evening in the nearest town, imbibing. The last bus home left before closing time, but it travelled via the only village between town and estate, and stopped at the Cushat's Nest pub just before last orders. Don was always first to nip off the bus, and made a swift entry to the pub. In his hand he held the correct money which he placed on the bar beside the whisky and pint that awaited him. With one sip and two gulps, the glasses were emptied, and Don joined the queue that was boarding the bus.

Alas, that was long ago; Don had aged, and worse . . . for months I had tried not to see it, and when I did to ignore it. But I could not, it was too obvious. Weariness had sagged Don's cheeks and dulled the sparkle in his eyes. His cheer had tarnished, and his body tired. Work became a struggle. I could avoid it no longer, and was forced to accept that my dear old friend was failing.

I suspect Don may have realised too, which would have explained a change in his character as he fought against it. At first it appeared to be only carelessness that caused Don to suffer a few minor accidents, but my alarm grew as I diagnosed something more unhealthy. Don had worked on sawbenches all his life and, all credit to him, his fingers were intact. He had paid the saw its due respect, but that was fading, and being replaced by the frightening sight of contempt.

*   *   *

An overnight gale had flattened a one-acre patch in Parson's Flatts. Unfortunately "windthrow" is an untidy nuisance, which had left us with a tangled mass of criss-crossed spruce to clear. Thankfully, that was almost finished and Don, Ron and I were to start the conversion of the small timber. Ron had inherited the Cundey peeler since I had been teamed with Don on the McConnel. I had no objections to the move because the sawing was more varied and interesting, and of course, I liked working with my old pal. Although Don was the official sawyer, the front end of a push-bench was hard work, and no place for my dear tired friend. When first paired with Don on the McConnel I had noticed his struggle and, under the pretext of wanting to learn to saw, I had relieved him of his job, relegating him to the back of the bench where the work was easier. And all had gone well until the first day's sawing in Parson's Flatts.

Ron had no shortage of props to peel; the smaller timber yielded a glut, but it left Don and me with piles of lengths just large enough to produce three-inch square posts from.

It was when I was taking great care to remove the thinnest of slabs from one of the smaller lengths that I experienced something I had not thought could occur. The end of the length had reached the tail of the saw when it suddenly stopped. Initially I did not know why, but I soon saw the reason. When a length of timber is pushed into a circular saw, the teeth at the front are rotating downward, and press the wood onto the table as they cut. If the timber is over halfway through the saw, and the cut nips, it grips against the rising tail section of the blade, which will throw the length up to the guard and onto forward moving teeth that fire it back at the sawyer. To help with this problem a riving knife is fitted immediately behind the tail of the saw. It is a curved metal blade, of a narrow wedge section that is slightly wider than the saw. Standing vertically just behind the rising teeth, it enters the cut, wedging it open as the length is pushed through the saw. A riving knife was, and still is, compulsory on all circular saws, and they are effective, but only after the timber reaches them. Before then, the cut can still nip and keep the sawyer awake. (See illustration on page 47.)

The slab I was cutting had been trying to nip and, being so thin, it had bowed outward as it was cut. Once past the tail of the saw the cut had closed at the end, and the leading edge of the slab had gone the wrong side of the riving knife. The cure was automatic: I had to pull the length back, and as I did so the rising teeth at the tail of the saw cut off the offending end of the slab. I only had to withdraw the timber an inch or two before I could push it forward again and complete the cut. It was only a minor hitch with a simple solution, one of those unforeseeable, million to one chances . . . that was going to repeat itself.

I half expected the second occurrence, which followed

shortly after the first, but as we worked into the morning the odds lengthened. The third time is said to be lucky, but in our case, on that particular day, nothing could have been further from the truth. The forward motion suddenly halted, again because of that fluke, but before I had the chance to pull back, Don, for some inexplicable reason, shot his hand forward and tried to pull the slab aside . . . with his finger! Not only was it an unbelievably stupid thing to do, Don was also attempting the impossible – he could not have pulled the slab out. I knew the gap between the cutting points of those rising teeth and the front of the riving knife was almost exactly the same as the thickness of the index finger that was inserted between them.

I watched helpless and horrified, awaiting the inevitable. I dared not move – if I pulled the length back it would have brought Don's finger with it into the saw. I was powerless: I could do nothing but yell – and that I had done from the first terrifying moment. I had begged, yelled, cursed, pleaded, shouted and implored for what seemed like an hour, but I knew I had wasted my breath when Don's face screwed up and I heard the new, sickening sound of teeth cutting bone, before he jerked his hand away from the tail of the saw.

At last it was over, and with the dread of expectation gone my emotions changed. I had just spent an age uncomfortably locked in impotence, watching hopelessly as my highly esteemed old friend deliberately perpetrated the act of an idiot: an act that lost him my respect and the end of one of his fingers. I was disappointed, annoyed: and then my temper flared!

Helpless no longer, I could go to the aid of my silly old pal and explain, with carefully chosen words, how impressed I was by his deed! I distinctly remember my body shaking to the point where it almost rattled with anger as I bandaged Don's shortened finger. I was seething, and as I tended the wound I blasted Don (and Northumberland, I was told) with the longest and loudest bollicking in the history of the county! I cannot print the monologue because of the fire risk, and the fact that the old fool laughed all the way through did not help! – Another waste of words that left a dense blue cloud shrouding the in-appropriately named Parson's Flatts!

The circular saw

Two days later it was Ron's turn. Will had replaced the injured Don and, as lunch time brought a break, I cut off the supply of diesel to the McConnel and Ron did the same with the Cundey. My machine stopped quickly, but the peeler did not. The Cundey had a single-cylinder engine and a heavy flywheel, and, coupled with the added momentum of the weighty blade-carrying disc of steel, it took a while before it came to rest. I wanted to disbelieve my eyes as Ron, after flicking off the fuel lever, immediately walked to the front of the machine and thrust his hand into the chute to remove some shavings long before rotation ceased. The lunatic, I thought, as I watched from 20 yards distant. The shavings were blown out of the spout, but some invariably collected on an internal corner that came within an inch of the draught-creating fans.

"Get your hand out of there!" I yelled, a second before I heard the sound of the fans contacting Ron's fingers. Why do I have to suffer these idiots, I asked myself, as Ron reeled away from the peeler, limping and clutching his hand.

"Quick, get the first-aid kit! Get the first-aid kit!" he howled.

"Let me have a look," I asked.

"No! No! I need the first-aid kit!" pleaded Ron as he hobbled, wincing, his movements contorted with pain.

"Not until I've seen your hand!" I ordered, unswayed by the demonstration. Ron, by then, had detected a lack of sympathy in my voice, and revealed his hand to show a slightly blackened nail – poor man! Two breath-taking, imagination-defeating acts of sheer stupidity in one week were more than enough for me!

*   *   *

Jeff arrived with our pay on Friday, and as he handed the envelopes out he mentioned that there was talk of a new sawmill being built, in a sensible place, on top of the hill above the old one. It would be equipped with the most modern machinery and would work regularly, and as Don was nearing retirement I was to be the sawyer – that was why I had been moved to the McConnel.

"But I don't want to be a sawyer," I objected. "I like working out in the woods."

"You seem to be happy enough on the front end of the McConnel," said Jeff.

"Yes; but it's outside. I don't want to be shut in a sawmill!"

"We can't have everything we want."

"I know, but if I'd wanted to be a sawyer I wouldn't be working on an estate. It's forestry and chain saw work that I want to do."

"Enough, son; you will do what you're bloody well told!"

A promotion I did not want – the thought nagged and lingered.

Etiquette

Doubt had rooted. My two years' practical were completed, and I should have applied for entry to a forester training school, but something stopped me. Contracting tempted me with its greater freedom and lure of more money, but, unsure, I decided to stay on the estate for another year.

Motorcycles were great fun, but did not protect from the cold and wet, and with a fourth winter approaching I had decided that four wheels with a roof would be better. After only six driving lessons I had taken my first test and had been failed, on a misunderstanding, by the most miserable examiner in town. He was notorious for his high failure rate and, according to my instructor, the only thing he happily passed was wind!

Of course, true to my luck, I got him again for my second test. This time I took the greatest care to impress my caution in one specific area, and the test seemed to go well. At the end I answered some easy questions – and then he came to the last one:

"When, Mr Surtees, does a pedestrian have the right of way at a pedestrian crossing?"

"When they are actually on it," I replied.

"Yes, that's right. I think you are rather overcautious at pedestrian crossings."

"I'm pleased you noticed," I said. "You failed me last time because you thought I wasn't."

I passed, and went back to work.

*　　*　　*

The past week had been particularly enjoyable. Unexpectedly I had been posted Next Door with the Jo Bu, and had been helping Tom clear some birch from a plantation of Norway that sheltered in a wandering, steep-sided valley. We had worked for a week and were into Saturday, when the sound of the hounds interrupted our work. Downing tools, we scrambled out of the wood and climbed the fence into the field to get a better view as the pack screamed into the bottom end of the wood. They ran well until a lack of scent halted them below us, and they were casting in silence when I heard a noise.

A tight wire on a fence can carry vibration a long way, and I thought the sound had come from the fence, but I could see nothing. It was repeated again and again together with a strange rasping sound. Looking along the fence I was not surprised to see a fox with its head and neck through the bottom two wires. With hounds casting within 30 yards, I would not have blamed it for leaving, but it did not. It seemed to be having difficulty making up its mind which way to go: its head and neck disappeared, reappeared, went back and came through again, each time twanging the wire on the fence, and making that other peculiar sound – very odd.

"What's wrong with it, Tom?"

"Don't know; looks like it can't get through the wires."

"But what's that noise? Let's take a look."

When it eventually saw us coming, the fox shot back out of sight into the wood, but we could still hear the unusual noise it was making, and when we came upon it we saw why. It was a small vixen pathetically tethered by a rabbit snare tightly noosed around her neck, and the strange sound was her gasping for breath. Foxes were numerous and pests, but I could not help feeling some sympathy for the hapless animal that cowered, entrapped before us. She had stretched the hounds at a pace, leaving a good scent for them to follow, but when she finally managed to check them, giving her time to gain ground, she had run straight into a snare that was not intended for her. Yes, I felt sorry for her – she had my kind of luck and I could not leave her to her fate.

"We'll have to get her out, Tom."

"Yes," he agreed. "It wouldn't be fair to leave her."

I looked for a suitable stick and, quickly finding one, handed it to Tom.

"You keep her interested with that, and I'll go round behind her."

Tom slowly pushed the stick toward the stricken animal, while I made my way to her rear. Thankfully, Vicky, as I called her, did as I had hoped and clamped her jaws around the threatening end of the stick that Tom proffered. Although fighting for breath, the vixen was energised by the adrenalin that self-preservation pumps into the system. The sight of those flashing teeth closing onto the piece of wood, that was an essential part of the rescue plan, reminded me of the danger.

"Right, Tom! Make sure she won't let go of that stick, and push her back against the snare so that she can't turn and bite."

I grasped poor Vicky in a no-nonsense grip below the snare around her neck, loosened the wire and carefully eased it over her ears. Once it was past her eyes, I could pull her back, sliding the snare along her muzzle onto the stick she now held loosely. She made no attempt to retaliate, and seemed to have given up although reprieved from a death sentence.

I climbed the fence holding the limp fox at arm's length by the scruff of her neck. She showed no sign of appreciation as I carried her along a hedge that pointed away from the wood and to safety. I took her a good 50 yards to break the scent, and as I put her down I hoped that someday she might realise what we had done for her. Vicky faced freedom as I released my grip on her scruff – whereupon she promptly demonstrated her gratitude by turning round, shooting between my legs and running straight back into the wood! The hounds were still there but Vicky escaped, it was her lucky day, but I often wonder what *she* thought.

*   *   *

My posting Next Door lasted less than a fortnight, but I enjoyed working with Tom again and we had the pleasure of Joe's company when he called one lunch time.

Joe's visits were usually unexpected, and always educational. His tales never failed to interest me and, by now, I had a much better understanding of the gulf that separated the life of a contractor from that of an estate woodman.

Our work was regular and supplied by one estate. The wages, except for the fluctuating overtime, were predictable. A cooked meal awaited us on our return home, and later, a hot bath would remove the day's grime and layers of rosin . . . Not so for Joe.

A contractor had to find his own work, and his income was at risk from the weather and other variables. And, if the travelling distance was prohibitive it meant staying at lodgings – or worse . . .

In short, a contractor lived a rough life, fogged with uncertainty; the work was hard, but Joe ensured that the play compensated – handsomely! By comparison, we on the estate were a sober bunch in steady, easy jobs, and the difference was highlighted in a story Joe told that lunch time.

Joe had just returned from Scotland, where he had been working for a month. Being prudent, Joe and the two friends he had worked with had opted to share a caravan parked on the job. Their temporary accommodation was well-equipped with a gas cooker (complete with a genuine frying pan); a unique ventilation system which harnessed the free power of any wind to exchange the air through strategically placed cracks, holes and other vents, and cold running water in the stream outside . . . Sheer luxury!

Fortunately, however, that small burn provided more than water: it yielded an unending supply of trout which helped to minimise expenses. The occasional pheasant also contributed. With the picture clear in my mind I could grasp Joe's reluctance to squander his hard-won money on non-essentials – like food – and appreciate the wisdom of his choice to invest it, instead, in life's necessities: alcohol, tobacco and women!

Both Tom and I listened intently as the tale continued; we could sense an imminent account of one of Joe's interesting experiences, and we were not to be disappointed. The tale centred on "etiquette".

My dictionary simply defines the word as "manners". However, we are all allowed our individual interpretation of the meaning of a word. To me "etiquette" means more than "manners". "Manners" are for everyday use, but I see "etiquette" as "advanced level manners", reserved for special occasions of pomp or dignity.

Joe and his two mates usually returned home each weekend but on this occasion they decided to remain north of the border to sample Saturday's nightlife in the nearest town. They were all in their twenties, single, and accomplished drinkers, with a special interest in the fairer sex. Gallant and with rugged good looks, Joe appealed to females, and it was in the dimly lit lounge bar that the trio met with three older, but very attractive, ladies. The introductions were soon over and the night began in earnest. They paired off quickly, and it soon became clear that they were all out for the same reason — to have fun! An endless supply of drink was quaffed to the sound of laughter, and intimacy developed as the alcohol worked. By closing time the final act of the evening was a foregone conclusion, and Joe was an excited man as his giggly partner led him out through the door . . . and into romance. The setting was idyllic.

Overhead, the moon shone brightly behind an obscuring blanket of cloud, as Joe's escort guided him into a dark back-alley. Slowly, they walked, and she held Joe tightly as he stood — and slipped — on a discarded bag of fish and chips. They cuddled, breathing heavily of a night-air freshened with the delicate bouquet from the glue factory. On they strolled, and then came the magic as their lips met — when the night was suddenly alive with the melodic skirl of two cats fighting in the nearby abattoir yard. And, as they continued to walk, she whispered lovingly into Joe's ear . . . She whispered words of encouragement to a man who had stood the pain of expectancy for over an hour! . . . At last she stopped, and pulled Joe into the blackness of an empty garage . . . She leaned against the wall . . . Now it was time for love!

A shower of stars lit Joe's darkness as the open palm of a

right hand delivered a stunning blow to his cheek.

"Och hell! De ye no hove ony etiquette, laddie?"

"Etiquette," spluttered a shaken Joe . . . "Etiquette?"

"Aye, laddie – etiquette! De ye no ken it's tits first!"

I still smile every time I hear that word.

Time to Move on

There was warmth in the air that blew onto me as I travelled to work; it came from the heater of my grey mini-van which was brand-new. I had savoured the luxury of being cocooned against the cold, wet and wind for the past month. Whereas I appreciated the comfort, I did not enjoy the drive this particular Monday morning because my destination was the nursery. Lifting time came with the back end of the year, and we had at least a week of the best job in hell to look forward to.

Time had brought change, and the change had made time increasingly precious over my three years on the estate. Hours were no longer measured in minutes and seconds, but in the new values of pounds, shillings and pence. With pressure pushing every job, Tom and John had been detailed to help with the lifting, but Don was missing.

My haggard pal had succumbed to "old man's disease", and was due to have an operation on his prostate gland that morning. Jeff had taken him to the hospital the night before, for the simple cure we hoped would restore Don to some of his former self. The routine operation was scheduled for ten o'clock, and at lunch time Tom went to see Don's wife for a report on his progress.

We had worked through five minutes of the afternoon when Tom walked back into the nursery. He wore a blank expression I had not seen before. I knew instantly . . . and hardly dared to ask. "What's wrong?"

"They haven't operated."

"Why not?"

"They started, and cut him open, but stitched him straight back up."

We stood, chilled with apprehension, as Tom took a long pause.

"The poor old sod's riddled with bowel cancer. He hasn't regained consciousness and isn't likely to. He's knackered."

A new and different silence fell on the nursery, and, gutted by grief, we worked as Don slept his life away. Mercifully, he did not suffer, and within four days my dear old friend was dead.

My sadness did not linger – anger burnt it. My old pal would never enjoy a single day of ease in retirement. He had toiled a weary two of the five additional years he had chosen to work so as to increase his pension by a miserable sixpence a week – an inducement that cheated him. No rest. No pension. No justice.

*   *   *

The estate was never to be the same again. My working life had already lost direction and the confusion increased for a while. Petty nuisances began to multiply into discontent in the blank left by Tom and Don. The nursery got even worse, and the thought of enforced "promotion" to the new sawyer's job niggled. My restlessness grew, and eventually I made up my mind. It was time to move on. Northumberland bristled with acres of trees waiting to be cropped. Out there, with my van and a chain saw, a better living beckoned – and timber would be falling for it!

214